What is DIP?

A guide to Document Image Processing

Martin Parfett

British Library Cataloguing in Publication Data

Parfett, Martin
What is DIP? (Document Image Processing)
I. Title
005.369

ISBN 1–85554–165–3

First published in 1992 by:

NCC Blackwell Limited, 108 Cowley Road, Oxford OX4 1JF, England.

Editorial office: The National Computing Centre Limited, Oxford House, Oxford Road, Manchester M1 7ED, England.

Typeset in 11pt Times by Ponting-Green Publishing Services, Sunninghill, Berks.; and printed by Hobbs the Printers Ltd, Southampton, SO9 2UZ.

ISBN 1–85554–165–3

Acknowledgements

I am grateful to Simon Perkins for the contributions made to this publication both in respect of material and comment.

Contents

1

Introduction

The objective of this publication is to introduce the reader to an emerging technology – DIP or Document Image Processing. It is predicted that this technology will have a significant impact on companies and their use of IT in storing, retrieving and accessing information traditionally held in paper form.

A number of factors are contributing to the increasing interest in this technology, not least the need for companies and their staff to make swift decisions. It is no longer possible to wait for the required information to be found and possibly take days coming to a conclusion. Today's business world requires rapid responses and decision making if companies are to remain competitive. The problem is compounded by the fact that the amount of information available is increasing daily, as can be seen by the increasing amounts of paper and computer generated information in our offices.

Much of the information within companies is used by more than one person but very few companies have any efficient way of dealing with the flow of information between staff. To save time, more than likely they will resort to issuing vast numbers of paper copies and the paper handling, distribution and storage problems that this brings, can constitute quite a costly problem. The cost of processing paper cannot be ignored. Staff time and storage costs account for a substantial part of our office costs.

The basic problem we are faced with is that most of the information we use in the decision making process is still in paper form. Figures from IBM and AIIM (Association for Information and Image Management) in North America claim that 95 percent of our information is still in paper form. A study by NCC of over 1500 firms found that a similar figure was applicable in the UK. This was despite the heavy investment in office technology which tended to generate more paper rather than reducing the amount in use. DIP provides a means of managing paper based

tion and therefore offers the potential of bringing under control a much higher proportion of a company's total information base.

A study by Norton Nolan found that the use of computer systems to handle 5 percent of a company's information base represented on average a spend of some 4 – 5 percent of a company's net revenue. DIP offers the potential to handle a figure calculated at nearer 50 percent of the information base and this fact must be borne in mind when considering the cost of its introduction.

In the following chapters it is intended to introduce DIP to you, not just the technology but also some of the non-technical issues, such as the question of justification and how to prepare for DIP. Chapter 2 provides an historical perspective, considering some of the developments such as microfilm which have traditionally been used to tackle the storage of paper based information. It considers what is meant by a document image and describes in a simple way the component parts of a DIP system and the difference between the structured information generated by a computer or word processor and the information held as a pure image within a DIP system.

The introduction of DIP is not just about the technology although in many companies it is certainly driven forward by the IT department. Chapter 3 pays particular attention to a number of non-technical issues which must be included when considering DIP. They range from a list of pointers to help with the identification of potential applications, to advice on undertaking the daunting task of a document audit and the possible impact, both on staff and the organisation as a whole, of the introduction of DIP.

Developments such as DIP will only be taken on board if there are benefits. Such benefits are described in Chapter 4 which sets out tangible, functional and corporate benefits. Tangible benefits include cost savings, for example, filing cabinets and office space whilst functional benefits are more difficult to express in monetary terms as they include such issues as improvements in procedural efficiency and retrieval performance. The third group of benefits can be even more difficult to quantify as within corporate benefits are such topics as service improvements, competitive advantage and strategic gain.

In order to help in the difficult area of justification Chapter 5 looks at the traditional approach to justification and then considers a supplementary approach which might help in the case of DIP. It is based on the concept of information economics developed in America and is seen as a way of supplementing the Return on Investment approach, which makes it possible to carry the idea of corporate benefits forward in such a way that it can be expressed in terms of a value to a company.

Another fact which emerged from the NCC Document Handling Survey was that companies had a very limited perspective as to where DIP could be used. They perceived it very much as an archival tool and as a replacement for microfilm. Chapter 6 seeks to dispel this belief and contains descriptions of a range of application areas in which DIP is being used. They range from the traditional records management arena through to workflow and transaction systems and item processing. To add additional value, this chapter also contains eight brief case studies illustrating different application areas and based on different types of DIP systems. The studies look at different information handling problems and how DIP has helped in resolving them.

As mentioned earlier DIP is a generic title which covers a range of systems. Chapter 7 aims to clarify this situation by classifying DIP in different ways and looking at the different categories of systems within each classification. Systems are classified by size, whether they are centralised or distributed, and by level of integration.

DIP is not one single technology but rather a number of technologies brought together and controlled by some sophisticated software. Chapter 8 addresses the technologies of DIP and starts by outlining the various stages of the imaging process. It then considers in more detail the image capture process including scanners and optical character recognition through to storage and the use of optical technology.

The final chapter considers the future for DIP looking at it not as a technology in isolation but rather as part of a move towards total information management. Comment is made on a number of technology developments which will affect DIP, as well as a number of related developments which may need to be considered in any long term-plans.

2

An Overview and Historical Perspective

2.1 Introduction

The objectives of this chapter are two-fold and set the scene for the rest of the publication. The first objective is to look at the development of DIP and those technologies whose shortcomings lead to the rise in interest in this subject. The second half of the chapter will define the major terms which are commonly used in association with this technology and to examine the different labels which have been devised by the computer industry for what is essentially the same thing.

2.2 Historical Perspective

Paper has for centuries been used in the office environment and during this time people have always sought new and improved ways of handling such material. The introduction of the typewriter had an impact on the amount of paper used as more words could be stored on a page and the modern day office would not be complete without the ubiquitous filing cabinets. More recently microform technology has been used, and for some applications very successfully, to reduce the amount of storage space required and also to improve the retrieval process.

Although microfilm has resulted in vast savings in storage, its main application area is in archival storage where access is not required as frequently, or the time factor for retrieving the information is not critical. Every site that needs access to the microfilm information will also need their own copy and viewing equipment. Use of Computer Assisted Retrieval (CAR) has certainly helped in this area by using a computer and database facilities to automatically index the microfilm and subsequently assist in the retrieval process. However, CAR does not help in the transfer and control of the microfilm images from the storage site to the reader.

Despite the predictions for the paperless office with the introduction of office automation in the 1980s paper still remained and technology resulted in more rather than less paper being used. It is so easy to produce multiple copies on word processors and to take photocopies, such that the storage and retrieval problems increase rather than decrease. The use of technology to overcome the paper handling problem still eluded equipment developers.

It was Frederick Wang, at the time President of Wang Laboratories, who commented that there had been three previous revolutionary landmarks that have radically changed the marketplaces approach to the way in which 'IT' handles information. In the mid 1960s it was the introduction of the IBM system 36 and in the mid to late 1970s the introduction of word processing. The early 1980s saw the widespread adoption of the Personal Computer. DIP was seen by Frederick Wang as the fourth revolution and like the others it was going to take some time to realise its full momentum and potential.

Also, what should not be forgotten is that DIP is not one single technology but rather a number of different technologies brought together by some innovative software. The establishment of DIP as an accepted technology has only been brought about by developments and advancements in each of these areas.

As a technology applied to business problems, DIP is relatively new. It is not so long ago that the first, large and expensive systems were introduced into the finance and government areas. The technology of imaging is not so new and can be traced back like so many technology innovations to the work of the space program which used such technology to manipulate or enhance satellite images. Using image techniques in this way required a considerable level of processing power and the technology was only able to move into the business arena through a number of technology advances.

The key aspect to remember with DIP is that DIP data is very bulky as it operates on a bit level and not a byte level and that even in a compressed form an A4 page of text in an image form can occupy between 10 and 25 times more space than an equivalent A4 page in ASCII text form. This obviously has severe implications on processor power, network topologies and bandwidths, as well as storage and display equipment. In addition DIP required the use of sophisticated database software which could run on smaller systems.

As a result of these technology developments, DIP is now a viable solution to many document and information handling problems and is not just a technology for the wealthy companies. One particular technology breakthrough which is always closely associated with the development of DIP has been Optical Disk Technology and this topic is looked at in more detail in Chapter 8.

2.3 What do we mean by a Document Image?

A document can be defined as 'a hard copy representation of information' although today that definition may be expanded to ' a hard copy or electronic representation of information'. An image in the context of this publication is defined as 'to form a digital representation of a document'.

For many years office workers, records management staff, librarians and the like have been handling images on paper. Typically documents such as books, reports, trading documents, etc, were produced and imaged on to paper for the purpose of reading. Even with the advent of office technology in the 1980s a wide range of office documents were still produced in a paper image form, even when word processors became widely used. A document therefore consists of one or a number of page images and such documents would be traditionally stored in a folder which in turn would be stored in drawers within filing cabinets.

2.4 DIP – An Overview

So, what is Document Image Processing (DIP) and how does it differ from other information handling systems? Trying to describe what it is, is not made easier by the fact that the computer industry is managing to confuse the issue by using different labels for essentially the same thing.

For example the following names are used for what is essentially the same thing. These include:

- document imaging systems;
- document image processing;
- electronic document management systems;
- digital optical storage systems.

These are a few of the descriptions picked up when scanning through recent articles. Preference seems to be given to Document Image Processing 'DIP' for two reasons. First, the name has already gained some currency and secondly, it describes quite well what it actually does. At this point, I think it would be useful to describe in a little more detail, the nature of a DIP system and process, and what advantages it offers over the alternatives. A more detailed description will be given in Chapter 8.

The majority of the information we all use is in paper form. Only a very small percentage is kept in microform (film or fiche). These of course are images but they are photographic images, which can very efficiently store large amounts of information, can be retrieved, sometimes with the aid of a computer-based index system, and viewed on a specially designed terminal. An even smaller percentage of the information we use, is actually created with the help of a computer or word processor within our own domain and is thus held in an electronic form. The text created in this way is called *structured or coded information*, and it can be stored and retrieved by the computer; it can also be edited of course. Neither paper based information nor microfilm can be called structured information as it cannot be read by a computer, it cannot be edited or manipulated, it is 'pure image'.

DIP offers us a way to capture this 'unstructured' information and convert it to a form which makes it manageable within a computer system, and once done, this information can be accessed and manipulated in much the same way as other computer generated data.

DIP really comprises a number of linked 'sub-systems', each with its own functionality, linked together and controlled by a system controller or processor. Different types of configuration make it possible to 'tack' DIP modules on to existing systems, or to have a stand-alone system, or yet again to have stand-alone systems with gateways to mainframe systems. It's really 'horses for courses', and therefore possible for all eventualities to be met. Figure 2.1 shows the outline concepts of a DIP system. First, a scanning device, accepts the document you want to capture, and working on the same principle as a photocopier or facsimile machine, reads the light and shade areas formed by the characters or drawings on the page and converts them to what is called a 'raster bit stream'. This is a computer readable code, which can be re-formed on a computer screen to provide a pure representation or 'image' of the page that has been scanned.

More advanced scanners can incorporate OCR (Optical Character Recognition) or even ICR (Intelligent Character Recognition) within the process so that one can save coded characters which can subsequently be treated in the same way as any word processed document. At this stage the image is usually only in machine memory, but it can be viewed on a screen and validated – to make sure all the detail of the page is faithfully represented to your satisfaction. The screen or monitor is normally a 'high resolution' screen, as the image quality of a standard personal computer or word processor may not be sharp enough to reproduce the detail required.

Indexing would usually be done at this point and only upon completion would the image finally be written to a storage medium. The performance of indexing and retrieval software is a critical factor in the configuration of a DIP system. The storage medium is usually an optical disk, but it doesn't have to be. An image can be stored just as easily on magnetic disk or tape. However, because of the size of image files, optical disk is usually the first choice. Again, because of its size, one

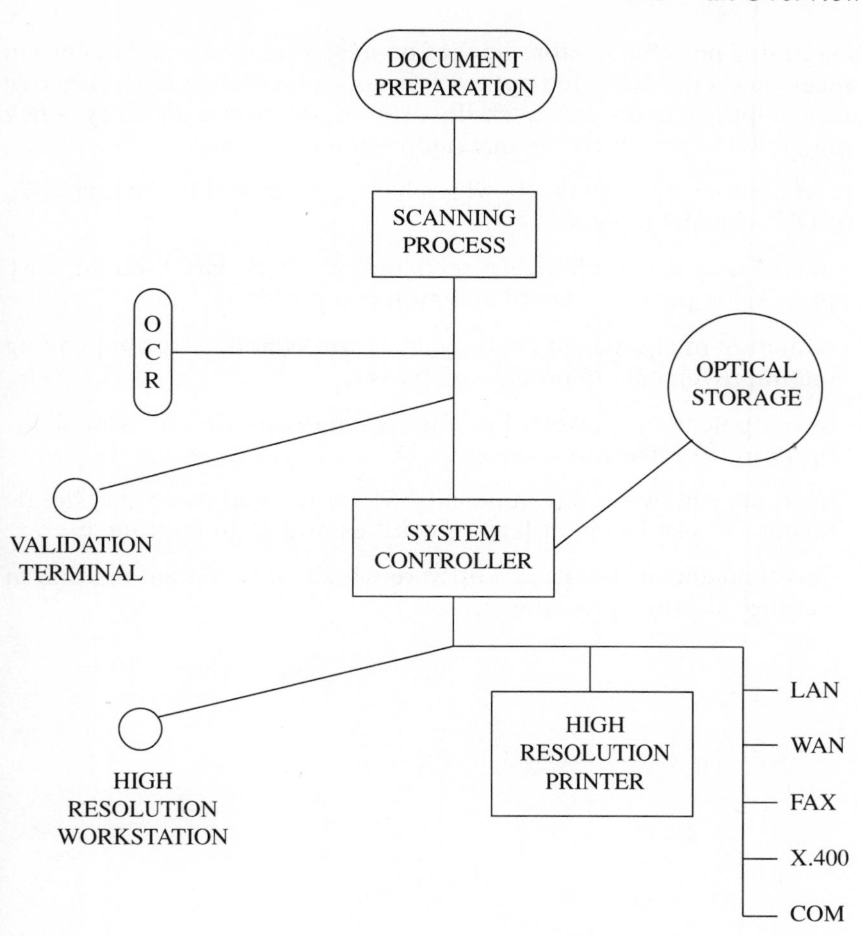

Figure 2.1 Major sub-systems of a typical DIP system

needs to compress the image to make the best use of the storage capability. This compression is simply a reduction of the number of bits required to represent the image, and often adopts existing standard algorithms used when transferring fax messages.

On the output side one would typically have a high-resolution monitor or workstation for viewing the retrieved image and a high-resolution printer for hard copy if required. Increasingly, the communications sub-system will include interfaces to facsimile as well as local and perhaps wide area networks. And herein lies the difference between DIP and all other alternatives, that once in digital form, that information can be accessed within seconds, simultaneously shared by a wide audience – in a secure environment – without any risk of violating or losing the original record.

Although it is possible to store images on magnetic media and in certain instances this is the preferred option, it is not a practical or cost-effective solution. Optical disks however do offer a solution especially when associated with improvements in compression techniques.

Other technology advancements which have contributed to the increased use of DIP systems include:

- developments in microchip technology which has brought vast processing power to micro and mini computers;
- reduction in equipment costs, such as workstations, accompanying the improvements in processing power;
- developments in powerful relational databases which allow flexibility in the retrieval of images;
- local area network developments which have allowed the simultaneous transmission of large quantities of data to multiple users;
- developments in workflow software which allow for automation of traditional office procedures.

3

Preparing for DIP

3.1 Introduction

When talking to potential users on the subject of DIP, discussions invariably turn to technical issues such as the effect of DIP on the company's communications network, or which document standards should be used when storing images. What most of these users have failed to understand is that the major issue which faces them is a need for a thorough understanding of the role and use of information which is flowing around their organisation. In this chapter we will look at a number of non-technical issues under the heading of preparing for DIP which every users must at some stage take into consideration.

All 'IT' implementations require careful planning but none more so than DIP. There is no such thing as a standard DIP configuration, each is usually built according to customer needs. As a result the planning and preparation for DIP tends to be less tolerant of omissions or oversights. Talking to existing users many had to contend with problems which despite careful planning still caught them by surprise. None of these problems were so catastrophic that they could not be resolved but they illustrated the need for careful planning because unlike other technologies, with DIP the source information is often destroyed once the paper documents have been scanned and so the DIP system is the only source.

Some of the problems users will have to be aware of include:

- it takes longer than expected. A reference to the fact that the installation of DIP should not be compared with installing a new financial package. DIP is a radically new technology and will need to be treated as such;
- new disciplines will have to be taken on board by management. In particular the analysis of information types and flows, and a structured approach to system planning are essential;
- the costs of and the time taken for, backfile conversion are often greater than expected;

– software inadequacies are frequent because the user's requirements have not been fully understood and specified. DIP systems are intolerant of such omissions.

With our current, predominantly paper-based systems we have preconceived ideas relating to storage requirements and the elapsed time needed to complete certain tasks and outdated procedures are often perpetuated. Not every DIP system will address these issues but it must be borne in mind that the objective of DIP is not just to provide an electronic replica of a paper-based system, but rather to looking at procedural and performance aspects and the way DIP can improve working practices and possibly contribute to a wider company objective.

3.2 Basic sources of information

By reading this book you have already begun your first task of finding out all you can about DIP, what it can do and what its limitations are. Sources of such information are not as widely available as you may think. Suppliers are an obvious starting point but there is no need I am sure to add a cautionary word about impartiality. You could join the UK AIIM (UK Association of Information and Image Management), details of which are contained in Appendix 4. AIIM is an independent forum for all those interested in information management problems and practices.

There are also a number of public seminars and conferences such as Optical Information Systems held annually in mid-summer although sometimes the range of different products displayed in related exhibitions often causes confusion. NCC also provide additional products such as the three part DIP Reference Volumes and the DIP Applications Handbook. Further details of these publications and other relevant NCC publications and information sources are contained in Appendix 3.

Also, do not dismiss consultants as they can and do provide a useful service especially where specialist knowledge is quickly needed. Buying-in the right kind of short-term help can certainly speed up an initial evaluation and possibly the later stages of implementation and still save time and money.

3.3 Identification of Possible Applications

Having obtained a broad understanding of DIP it will allow a prospective user to form an opinion about the relevance of DIP to their organisation. Once this has been achieved it is possible to identify the best, or most likely application areas. The following are a number of application pointers which may help in the identification process:

- high volumes of paper documents;
- documents originating outside the organisation;
- high-value documents;
- high cost of storage space;
- high cost of information retrieval;
- delays in information retrieval which affect customer service;
- exposure to risk due to non-availability of document;
- intensive manual document processing;
- lengthy transaction processing;
- processing delays resulting from physical locations of storage facilities.

Once you start looking at possible application areas it will become clear that DIP should not just be considered as a replacement for a filing cabinet. Certainly there will be occasions when DIP can be justified on the basis of providing a more cost-effective storage and retrieval facility. Increasingly however DIP is finding its way into more volatile business areas where the access to, and assembly of, disparate elements of information, where an improved service response is needed or to supplement a decision or course of action. It is also becoming a front runner in paper intensive transaction processing systems such as loan applications, insurance claims and invoice processing.

Most companies who have implemented DIP first opted for a small pilot scheme rather than adopting the new technology on a corporate-wide basis from day one. This approach helps both management and staff come to grips with the new technology and also the cultural changes that it will impose. The removal of paper-based information on which we have all relied for so many years calls for very careful preparation.

3.4 Beginning the Process

The manager of one of the first DIP installations in the UK said that the best advice he could give prospective users was that "they must comprehensively understand their paper process in precise detail before letting DIP anywhere near it". This statement was based on practical experience of the problems that can result if there is too much of a rush to install the technology before the underlying processes are fully understood.

It is at this stage in preparing for DIP that some good old fashioned O & M skills come into their own as the first practical step in the process is to carry out a thorough document audit which should determine a number of points. These include:

- the volumes of all documents whether typed, printed or hand-written;
- the retrieval patterns of all documents;
- whether they are internally or externally generated;
- the physical condition of the documents (ie are they folded or the contents fading);
- the life cycle of the documents.

It is a daunting task to undertake but it is required if you are to get the best out of your system.

The next step is to find out who uses, or may have occasion to use the information in question. The exercise should include those people who keep duplicates of all or part of the information held by the prime work group being studied. It is also necessary to determine the internal and external relationships between the work group and other departments, or between the work group and customers or suppliers. Such relationships should be closely examined to determine if they require changes or improvements. The objective is to build up a comprehensive picture of information use and flow throughout its life cycle. To complete the picture it is necessary to have a clear understanding as to how the work of the group under investigation contributes to overall business objectives. Such an exercise serves two purposes. First, to be able to build an effect system requires a comprehensive understanding of the work groups' problems, needs and objectives and secondly, much of this information will be needed in the justification process.

Once bitten by the DIP bug it is very easy to assume that it is the only answer to your document handling problems. Despite all the recent press comment on the possibility of the paperless office at last becoming a reality, and despite what the product vendors may say, there are still occasions when cheaper alternatives can still provide the most cost-effective solution yet still provide an upgrade path into DIP when this is justifiable. Even if you think DIP is the solution you must not close your mind to the alternatives. Senior management will expect that as part of the justification process all alternatives have been considered and costed.

The specification of requirements documents has an important role to play in preparing for DIP. As the system's infrastructure supporting corporate IT facilities increases in complexity, so too does the risk of shooting yourself in the foot through an oversight. Writing this statement of requirements will not necessarily guarantee that this will not happen, but it does add a further level of evaluation from the perspective of an operational rather than a theoretical system. The problem with this specification is that too many staff are more concerned about the size of the optical disk and the type of scanner that they want rather than specifying in detail the business problem they want to solve. Some suppliers like this approach but you might end up with the wrong solution if you provide just a technical document and could be held accountable by a supplier if you get it wrong. The specification also establishes some consistency for the briefing for suppliers and should be used as a base document against which to judge their responses. Although the document audit and information flow and analysis are time consuming tasks they are essential. The value of this task appreciates as the specification of requirements is prepared and justification process gets under way (see Chapter 5).

3.5 Information Transfer

When preparing for DIP one of the major issues which needs consideration is how to deal with the existing data contained within the current manual filing system. This can often be the sting in the tail of any implementation and can be expensive in terms of effort required and costs. In essence a company is faced with three choices. These are:

- they can run parallel manual and electronic systems with an agreed cut-off date. This approach results in considerable extra work, a possible disenchantment in the benefits of DIP and the need to have suitable skills to run two different systems;
- they can have a phased introduction, but this often results in confusion as to which system holds what information;
- they can consider backfile conversion. In other words they can address the issues of converting existing manual information into an electronic form and sooner or later this will be an issue with any DIP project.

It is the third of these options which will be looked at in more detail in the next four paragraphs.

Whatever happens backfile conversion will be very critical to a DIP implementation and it is not cheap. It must be carried out with minimal disruption to the normal workflow and this in its own right can cause problems.

The first step is to categorise the information which is being considered for conversion. This can be of two major types but with a third group which should be taken into consideration.

These are:

- archive information. This is information in the corporate domain which as the name suggests is not active, but is held as an archived facility. From the point of view of causing the minimum disruption of normal work, this category of information is easy to deal with;
- transaction information on the other hand is more difficult as this is the information that is being used as part of the normal office procedure. In other words, it is the insurance claim that is being processed or the customer correspondence that is being dealt with;
- personal information. We all have it, sitting in the bottom draw of our desks. This is information which individuals have collected and possibly nobody else is aware of but which is of considerable help in our day-to-day operations.

The options we have when dealing with the backfile issues are either to deal with it in-house or look to one of the bureaux that are now available which offer such services. To give an example of the amount of effort required, one company undertook the backfile conversion of 80K documents and carried out a 100 percent quality check. This required some 30 man weeks of effort to carry out the work. For the remaining 300K documents, they only undertook a 10 percent quality check and this required 36 man weeks of effort.

If the work is undertaken in-house, there is a better sense of control as you know your own business and its possible day-to-day needs, but it will require the purchase or rental of extra staff and equipment in the short term and mistakes and delays can be expensive. If you go for the external solution, you should find experienced and specialist staff and the quality and speed of capture will be much better, but the cost will be considerably higher.

However before this decision is taken a second and more detailed audit of documents is required. This should entail the following steps.

- identify the different categories of documents/information;
- analysis of documents by volumes/process/lifecycle/physical condition, ie torn or folded, etc;
- carry out a benefit analysis to determine which documents would be most beneficial to convert first;
- consider processing alternatives. This is important for DIP and should cover such issues as whether the document can be indexed as a batch of documents rather than individually;

- each step of the conversion process should be spelt out and sequenced;
- a management guide setting down responsibilities, timescales, etc should be produced.

3.6 Preparing People

As well as needing careful planning in the information transfer stage, the issues of preparing staff for DIP should not be forgotten, and compared to the introduction of other office technologies DIP will have a profound affect on the way that a company can operate. We have all been used to working in a paper-intensive environment and the introduction of DIP will require a radical cultural change to working practices and procedures.

Staff should have a measured involvement in the planning process and there is a need to break down the dependency on paper. We need to move away from the typical approach to electronic mail that once a message has been received, a hard copy printout is produced. Training and education of staff is so important and the needs of different age groups and different levels of staff must be taken into consideration. New skills are required, such as indexing and scanning, and companies must not allow their familiarity with new technology to blur their training needs analysis.

3.7 Potential Impact on Organisations

As well as affecting the staff, the very organisation itself can be affected by the installation of DIP. It presents an opportunity to examine and, more than likely, revise many existing practices and procedures. Such changes are essential if the full benefits of this technology are to be realised. New skills will mean that the traditional job specification for office staff will have to change and skills such as scanning, indexing and electronic retrieval will have to be included.

In North America we have seen examples of how the use of DIP has started to break down the traditional ranks and hierarchies of middle management. They are no longer required to control the processing and evaluation of information, as many checks and prompts can be built-in to the systems. This development may be slow to materialise in a more conservative European business environment.

One final effect on the organisation that needs to be mentioned results from the fact that through the use of DIP, companies are now in a position to be increasingly more responsive to fast changing needs. No longer can the excuse be made that it will take three days to find all the information. With DIP such information is more readily available and companies will need to make use of this fact.

3.8 Security and Legal Issues

Both Security and Legal issues will be of concern to the organisation and as part of the planning process due consideration of them must be taken. From a security point of view, the threats that need to be considered include:

- fraud;
- mischievous hackers;
- malicious damage.

Like any new technology development, and I can think of examples with the introduction of EDI, companies take advantage of the uncertainty in this area, to slow down or even stop the uptake of technology, but careful planning and thought will overcome most of the problems.

From a corporate point of view security and legal requirements must be determined, hence the need at some point to include your legal and audit staff in project meetings. This corporate view can then be interpreted at an operational level. There is increasing support for the use of DIP, as electronically stored information need be no less secure than paper and although 'WORM' (Write Once, Read Many Times) technology can be fraudulently used, so can many of our existing systems. It is important that you check legal acceptability with information users so, for example, if you intend to keep invoices on a DIP system, check the position with HM Customs and Excise.

3.9 Conclusion

DIP is seen as a very technical subject but it impacts on the very fundamentals of the way that systems or procedures have operated for years. As such, there are many non-technical issues which have to be considered and dealt with. If there are two key messages that I would like to leave with the reader from this chapter, they are first, that the *analysis* and *planning* of a DIP installation are vital. If you skim over these time-consuming task it will only rebound in the long term, as companies have seen to their costs. The second message is do not underestimate the implementation time required for a successful project. Such underestimation can be fatal.

4

The Business Benefits

4.1 Introduction

When establishing the potential benefits that DIP could bring to an organisation it is important to understand the dependence that all companies have upon information in its many forms. Traditionally this has been a relatively straightforward operation and much of the information gathered was used to confirm, or otherwise, the correctness of past actions. Little effort was extended in gathering external information and most of that used came from business correspondence, newspapers, professional journals and the like. Much of the internally generated data met the needs of financial management for number crunching and monitoring of sales, income, costs and production levels. Middle management made extensive use of such information and most of their decisions and recommendations to senior management were based on it. However such information suffers from two major problems both of which impact on the decisions being taken and the recommendations being made. The information was historical and it was very much a parochial view.

Although technology in the last five years has moved on at a rapid pace, presenting us with such tools as executive information systems, sophisticated decision support tools and other productivity aids, which certainly allow managers to process and evaluate large amounts of information very quickly, the fact remains that much of the business information is still held in paper form and is outside the electronic environment.

Paper-based information is ideal for archival purposes but it cannot be retrieved or searched using computer technology. Despite these major problems it still remains the principal storage medium for information. The amount of information coming into any organisation or needing to be used by the organisation is forever increasing. Examples of some of the areas contributing to this situation include:

- increased government legislation:

    - the Data Protection Act
    - Health and Safety at Work Act

- the need for better business information:

    - competing products and price structures
    - new markets
    - the effect of government policy on business

- environmental issues;

- the European Community:

    - new legislation
    - the Single Market

- increased use of high technology:

    - advanced production machinery data
    - information technology.

You only have to look at the daily post arriving on your desk to see that the possible sources of information are also increasing and what we see as today's information handling problem will get worse and a company's ability to manage information as a resource will become very significant if the business is to succeed. It is now said that information is a company's most valuable resource although it does not seem that long ago that the very same thing was said about a company's staff!

Why, you may ask, has this chapter started with a look at the role of information and the problems of information overload? Simply, it is to illustrate that strategically DIP should not be looked upon in isolation. It is a contributor to solving a much wider information handling problem. Those companies who look at DIP as just one piece of the jigsaw rather than as an end in itself will ultimately receive the most benefit.

If DIP is so important for achieving a long-term solution to our problems what are the benefits that can be expected? In the rest of this chapter three different areas of benefits will be outlined. These are;

- tangible benefits;
- functional benefits;
- corporate benefits.

4.2 Tangible Benefits

It is in the area of tangible benefits that many of the first savings were made using DIP. Such areas are usually easily identified as they represent an operational overhead.

4.2.1 Hard Copy File Storage and Floor Space

The contents of five or six four-drawer filing cabinets, which seem to fill every modern office, could be stored on one 12" optical disk. Such a statement depends on certain assumptions, but it does serve to indicate the potential savings in floor space that can be made. Obviously, the more expensive the floor space the filing cabinets are taking up, the easier the justification process is.

There are numerous examples of companies in high-cost locations who have recovered their DIP investment costs within two years for such an application. What has also become evident and can be seen as a spin off benefit, even if DIP is not introduced ultimately, is that in the identification and evaluation of possible applications changes and improvements in existing systems and procedures have been identified and implemented.

4.2.2 Savings in Photocopying Costs and the Cost of Paper Copy Distribution

Photocopying facilities are normally available on an open access basis with little or no control exercised on usage and cost reallocation. What should not be forgotten is that this approach can be expensive and when one takes into account the cost of staff time both clerical and professional, the paper and usage charges plus the additional storage costs for all the thousands of copies taken for internal use, this can contribute a significant amount towards the system justification.

Linked to the photocopying issue is the subsequent need to distribute resultant paper copies of documents. This can be both using an internal mail system or via a public postal service.

4.2.3 Time Spent on File Maintenance and Retrieval

The cost of running a paper filing system is normally not known and therefore it is treated as an acceptable company overhead. From the experience of those companies who have tried to establish the true cost, the figures are higher than expected and should be used in the return on investment calculation.

DIP can also eliminate the problems of misfiled papers, damaged or lost originals and files out with other users, all of which are causes of lost time when looking for information in a paper-based system. Use of DIP can also mean information held in valuable and rare documents/books, to which access is normally restricted, can now become more widely and readily available.

4.2.4 Cost Avoidance on Headcount and Accommodation

The introduction of the poll tax in the UK with the resultant huge increase in supportive documentation, proved a headache for many local authorities in terms of the extra staff required to handle these increases and the subsequent need for additional accommodation. In some authorities however, the introduction of DIP enabled existing staff after suitable training, to deal with the increased workload with little if any extra space required. A related benefit is the improved staff moral resulting from the use of 'leading edge' technology.

4.3 Functional Benefits

Trying to put a value on functional benefits is especially difficult when there is pressure to express such benefits in monetary terms. The only real yardstick against which to measure is improvements in productivity and it is important to clearly identify problems which are to be addressed, and the quantifiable improvements that are being sought.

4.3.1 Procedural Efficiency

Improvements in procedural efficiency come under a number of headings:

- improved staff productivity;
- availability of information;
- retrieval performance;
- monitoring and control.

Many DIP implementations report improvements in staff productivity, improvements which can be both clerical and professional. Although the clerical savings can be substantial, the possible savings relating to management and professional staff should not be forgotten. There are many occasions when, due to past experience and knowledge, it is the professional staff who end up looking for information. In the legal profession it was not unknown for senior staff to spend hours looking for documents to support a case. The use of DIP has reduced the retrieval time down to seconds and resulted in clerical staff doing the work, thus freeing the legal expert to do the work he is paid for. In some cases DIP is seen as a way of dealing with anticipated growth rather than employing more staff.

When considering the advantages that DIP has over alternative technologies in the area of information availability there are a number of factors in favour of DIP. One major factor is the ability to provide for the information retrieval needs of multiple users from a single source. Although microfilm has provided major benefits and savings over paper based systems it has a number of disadvantages. These include:

- no support for remote access;
- no support for full text retrieval;
- the need for multiple film copies;
- the need for multiple viewing terminals;
- no support for multiple simultaneous access.

Even with proficient clerical and administration staff, manual filing systems still require several minutes of user effort to obtain the required information. The use of powerful indexing and retrieval tools associated with DIP can dramatically improve retrieval performance.

DIP also offers benefits in the use of workflow software to improve the control and monitoring of many office procedures. In the same way that we have automated many of the blue collar tasks over the last thirty years it is now possible to automate many white collar tasks. More information on Workflow Software is given in Chapter 8. With this additional control it is also possible to provide a complete audit trail with detailed statistics and reports on operator and system performance.

4.3.2 File Management

The maintenance of paper files has always been a problem but more recently the problems of finding suitable conscientious staff to carry out this task has aggravated the situation. On many occasions papers get misfiled or filing is left undone and consequently the integrity of information held in such files is brought into question.

In industries, such as oil and nuclear power, staff must **always** be able to rely on the integrity of their information and this also applies in many commercial operations. DIP can therefore provide substantial benefits in the area of file management but there are still two areas, document capture and indexing, which can have a high manual input, and which can impact on the integrity of the file.

4.4 Corporate Benefits

When undertaking a return on investment calculation in respect of a proposed DIP installation, the direct benefits which can be made are often cancelled out by the high cost of equipment and there is a need to look to other areas of benefit to make the business case. With DIP, many of the additional benefits can be classified as corporate benefits, which are very difficult to define this and is why in Chapter 5 consideration is given to the use of Information Economics as a way of putting a value figure on such benefits.

There is a growing acceptance of the concept of corporate benefits such as improvements in the quality and timeliness of management information. Other benefits in this category include:

- competitive advantage – a much used phrase, but which was claimed for many early installations of DIP systems in the financial sector, where the ability to process an insurance claim or a mortgage application in half the time of a competitor, does indeed provide competitive advantage;
- service improvements – there are now numerous examples such as the Severn Trent Water Authority where DIP is being used to provide an improved level of service to customers. All relevant information can be made available on a single screen to a customer services clerk enabling a level of response that today's client base expects;
- perceived image of excellence – use of DIP within a company is seen as a way of projecting an image of excellence. This has helped to recruit quality staff and also to improve customer relationships;
- strategic gain – this is very difficult to define and can vary between organisations. One definition used of such a corporate benefit is "an influence which goes beyond meeting immediate operational objectives and which can positively impact organisation structure and/or direction and therefore performance".
- value added benefits – these are benefits were the use of DIP enhances other parts of the business. Examples include:
  - through the use of DIP and its integration with other IT systems, transaction documentation is no longer isolated from the rest of the information services;
  - certain functions could now be decentralised and be carried out in less expensive office space.

Experience to date has shown that in certain cases the functional and tangible benefits are sufficient to justify investment in DIP. However many of the successful implementations see DIP very much as a strategic development and one which can contribute to the achievement of corporate goals; it is in this area that the major benefits are expected.

5

The Justification Process

5.1 Introduction

Within the UK, the process of justifying DIP has been a thorn in the side of those staff who have identified the business benefits that the technology can bring. Having seen an opportunity to improve the efficiency of white collar staff or to improve the service offered by their organisation they then have the thankless chore of persuading the senior management team to fund the project. Not all UK companies take this approach and there are numerous examples of DIP pilots, costing £50–100 K for hardware alone, being implemented without a formal Return Of Investment (ROI) justification exercise taking place, simply because of the company's confidence in the potential impact that this technology can bring.

Within the rest of Europe the approach to justifying the investment in DIP differs. For example, in Germany improvement in clerical productivity is the key to justification, whilst in Italy the benefits that DIP can bring in the archival area are more accepted. In France, there would appear to be a more liberal approach to the justification process; users want to include image capabilities as part of their office automation facilities and cost justification does not appear to be the barrier that it is in the UK. In the UK there is a much greater emphasis on hard savings either in respect of people or storage space.

5.2 The Traditional Approach to Justification

The provision of computing services has for too long been controlled by the finance function and the systems and services provided have been largely aimed at middle and senior managers. Such systems have tended to concentrate on reporting on past performance, collecting and collating information from operational levels of the company and providing statistical performance factors which help in the control and management of the business. In such an environment the introduction of any new system has often had to be justified on a traditional Return On Investment calculation, ie new computer system is equal to a reduction of x number of staff with a pay-back over three years.

However, recent developments have seen a move away from the monitoring and control with new systems, making it possible to improve efficiency at operational levels. The Personal Computer has played a major role in this shift of emphasis, which has put computing power in the hands of the end users. In many cases PCs, faxes, mobile phones, etc have been introduced with very little thought of justification.

It can be argued, therefore that because of the rate of spend in the last five years on end-user computing, with little apparent return on investment, that putting a case forward for investment in more expensive technology to help the users control paper-based information will meet with little support. If the case for DIP is to be sold successfully it must be aligned with the business. Typical business related goals include:

- improvements to customer service;
- the enhancement of company image;
- improvements in internal white collar productivity (both managerial and clerical);
- tangible reduction of costs whilst maintaining or improving facilities.

The problem comes when trying to quantify the degree of benefit that can be achieved when taking this approach and in many cases innovative technology projects are competing for limited capital funding alongside projects from manufacturing, marketing or sales. As senior management can relate to those other traditional projects, it is even more difficult to sell the fact that new technology can contribute to the health and growth of the business and this is crucial for the justification process.

Cost benefit analysis has long been used to justify the acquisition of new technology. As a way of arriving at a statement of return on investment (ROI), cost benefit analysis can tend to mask the true business benefits behind a sea of numbers and calculations. Anyone who has had to put a justification case forward will know how figures can be manipulated to prove anything you want. The same figures can show a good or bad case, depending on the case you want to put forward.

As a starting point, the ROI calculation should be undertaken as it is a well recognised and accepted business practice and it is one which our accountants and finance managers are accustomed to. In North America where DIP has been more accepted the approach to justification is not so fraught. Management have a far greater acceptance of the value of access to information and its contribution to business goals but in the UK we still have to put forward a set of thoroughly researched and cogent arguments.

5.3 The Process of Justification

The following is a brief overview of the justification process which should be followed when introducing DIP.

5.3.1 Understand the Problem

This basic issue is often overlooked or skimmed over and misunderstandings can lead to problems later in the justification exercise. Often this requires an understanding of the different information types and knowledge of the way information flows through an organisation.

5.3.2 Understanding the Objectives

Before any solutions can be considered one must have a clear understanding of why improvements are being sought and what are the required end results. There may be different objectives ranging from a need to make savings on expensive storage space and a desire to centralise an information resource, to eliminate any loss of information or the integration of paper-based information with other existing data systems.

5.3.3 Identify and Evaluate Alternatives

DIP is a technology which can so easily impress potential users and it is very easy to assume that DIP is the correct solution. It may not be and part of the justification process is to consider other solutions such as improving a paper based system or microfilm rather than just assuming DIP is the answer. Managers will expect alternatives to be considered as part of the justification process. The evaluation of alternatives should consider the performance elements of any option as well as cost implications under the following headings:

- capital costs (hardware, software, etc);
- ongoing costs (maintenance, insurance, etc);
- associated costs (training consultants);
- savings and benefits.

5.3.4 Justification Arguments

Presenting a series of justification arguments to senior management can be fraught with problems in normal 'IT' investment and in the case of DIP it can be even more difficult. DIP is different and should not be seen just as a cost-cutting technology and often there is a need to take a much wider view and assess the importance of using DIP for strategic gain. Some of the unique features of DIP are:

- it is not exclusively application specific;
- transferring paper based information to the screen will call for large changes in user attitudes;
- new skills will be used such as indexing;
- it is more costly and more sophisticated than other information handling methods;
- it will pervade all levels of an organisation.

The present justification arguments have a number of shortcomings which fail to take into account some of the special requirements of DIP. These shortcomings include the fact that:

- the traditional methods address them to satisfying operational need and fail to recognise the need to link any proposed performance improvements to line of business objectives;
- there is a gulf between perceived operational need and corporate directives and hence between operational and executive levels of an organisation;
- there is too strong an emphasis upon the need to justify through cost savings and little or no attention given to potential strategic gain.

5.4 A Supplementary Approach to Justification

Early in this chapter a number of reasons were put forward to show that DIP is different from other technologies and as such the traditional approach to justification may be inadequate. There is nothing in fact wrong with the traditional ROI approach and it provides a sound basis on which to develop. Cost benefit analysis helps to define the benefits we can expect to achieve and also their cost. It also considers the affect of cost displacement and reduction and, as mentioned earlier, this is an approach which our financial controllers are comfortable with.

However, using the concept of Information Economics, developed by two Americans, Parker and Benson, it is possible to carry the notion of corporate benefits forward in such a way that they can be expressed in terms of a value to the business.

It is not possible in an introductory publication such as this to go into any great detail on the concepts of Information Economics. Readers who find the subject of interest and require more information should refer to the *NCC DIP Reference Volumes* and, in particular, the work done by Simon Perkins on justification.

Parker and Benson stress in their approach the need for a two-domain analysis which will show the relationships between the business domain and the technical domain of any organisation. Figure 5.1 shows this relationship and demonstrates the need to view projects such as DIP from a wider perspective.

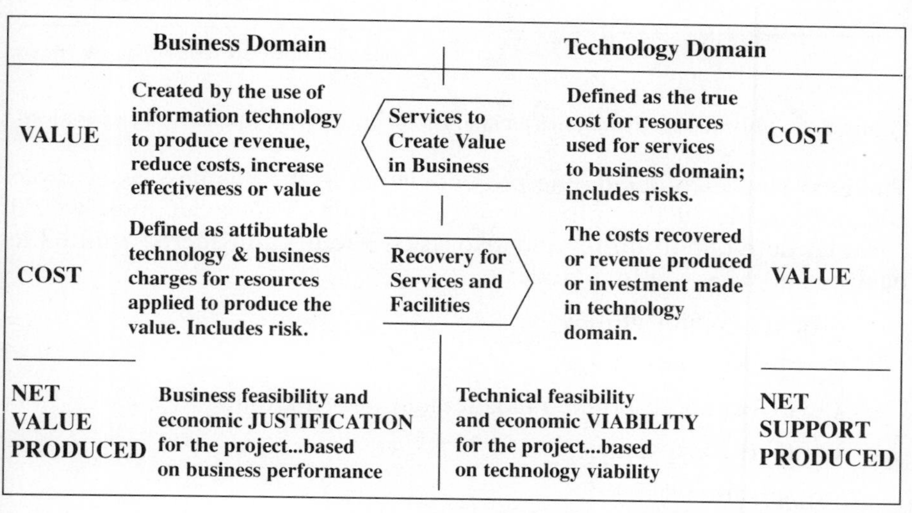

Figure 5.1 Information Economics two-domain Model

At the heart of this supplementary approach to justification are two fundamentals:

- the notions of value;
- the proper assessment of risk.

The success of the method depends upon a systematic presentation of influential factors, made to a balanced team of senior managers ideally representing all disciplines, who can then apply a structured assessment method to decide the worth of the DIP project to the company. The first step is to identify all possible elements of value which can be used to supplement the cost benefit exercise. Examples of some of these could include:

- will the project result in the earlier achievement of operational or financial goals?

– can 'spin-off' advantages be seen in other areas of the business as a result of implementing, say, a departmental project. Figure 5.2 shows how the ROI calculation can be developed using additional value judgements.

Traditional Cost–Benefit	+	Value Link-ing	+	Value Accel-eration	+	Value Restruc-turing	+	Innovation and investment Valuation	=	Input to Simple ROI Calculations

Source: Information Economics by Parker and Benson

Figure 5.2 Information Economics techniques for developing simple ROI calculations.

The next step is to decide the projects value to the business as a whole and not simply to the initial application area. This evaluation would consider both opportunities and also risks. Factors considered within the business domain would include:

- strategic opportunities;
- competitive advantage;
- the effect on corporate management information;
- competitive response;
- organisational risk.

Within the technical domain the following would be considered;

- the effect of DIP on the systems architecture;
- the clarity of project definition;
- the technical uncertainty or innovation resulting from the introduction of DIP;
- the effect on the technical infrastructure.

Figure 5.3 shows a typical score card used in the evaluation which could include both positive and negative scores. Using this approach the final weighted score can be used to evaluate the merit of a high-value IT project with another perhaps from say the production side of the company looking to install new machinery or a request for a new marketing campaign.

Information Economics must not be seen as an alternative approach to justification, it is a supplementary one. It is a concept which can be applied to any high-value information system when the majority of perceived benefits are in the corporate arena.

Evaluator	Business Domain						Technology Domain				**Weighted Score**
(factor →)	ROI* +	SM* +	CA* +	MI* +	CR* +	OR* -	SA* +	DU* -	TU* -	RI* -	
Business Domain											
Technology Domain											
Weighted Value											

***Where:**

ROI Measurement

ROI = Enhanced simple return on investment score

Business Domain Assessment

SM = Strategic match
CA = Competitive advantage
MI = Management information
CR = Competitive response
OR = Project or organizational risk

Technology Domain Assessment

SA = Strategic IS architecture
DU = Definitional uncertainty
TU = Technical uncertainty
IR = IS infrastructure risk

Source: Information Economics by Parker and Benson

Figure 5.3 A typical Information Economics score card

6

Application Areas and Case Study Material

6.1 Introduction

The objective of this chapter is to overcome one of the negative user perceptions of DIP, highlighted in an NCC Document Handling survey. This showed that there was a very narrow view on the range of applications for which DIP was suitable. The prevailing opinion was that the main application area was records management, perhaps reflecting the publicity given to early DIP implementation which were certainly of this type. However, like most technologies the cost/performance ratio has improved and with a number of innovative developments from users and suppliers alike, this technology is now being considered for a wide range of application.

6.2 Document Image Processing Applications

In general terms every DIP application will make use of the same components as shown within the basic model of DIP process featured in Figure 6.1. What is different is the degree to which every component is used and this is often related to document size and other factors relevant to the information. The document size is important because it impacts on the type of input and output peripherals, the workstation requirements and other basic system architectural considerations. Detailed below are a number of DIP application areas.

6.2.1 Records Management

Such applications are typified by the high volume of data input and the relatively low retrieval rates, hence a need for few retrieval terminals. They are used to save office space, to speed the retrieval time for requested files, etc and are often installed to replace existing paper and microfilm systems.

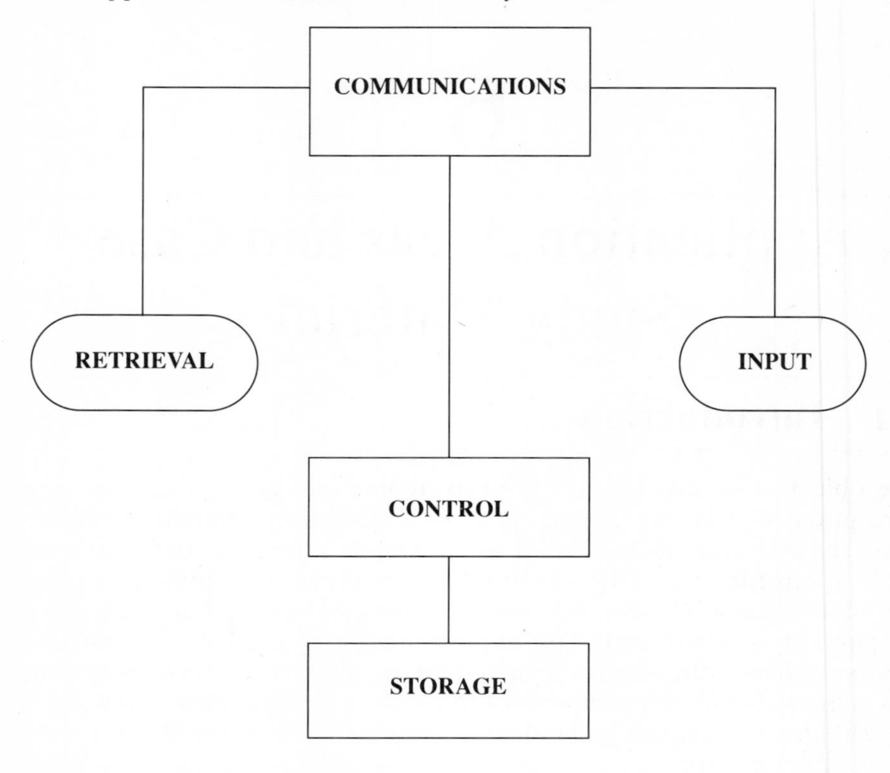

Figure 6.1 Basic model of DIP processes

In such applications there is a need to preserve the documents in page image form and the question of storing such documents in an agreed document standard such as ODA (*Open Document Architecture ISO 8613*) has to be considered. This is because the retention period for such information can be decades and there is the possibility that the stored images may need to be transferred from one storage system to another during its long lifetime. Application areas include libraries; central and departmental registers, project documentation centres and archive centres. The sheer size of such systems often results in a very high cost and alternative technologies such as microform may be strong contenders to DIP.

6.2.2 Engineering Document Management Systems (EDMS)

Such applications are often identified by the need for special requirements associated with the capturing and printing out of large-sized documents. Such systems often have to cater for AO sized drawings and also be able to interface with the 35mm roll microfilm and aperture cards which have traditionally been used in this application. Another size-related issue is the requirement to handle 'tiling', in which large bit-map images are divided into a series of tiles composing 512 x 512 pixels in each. Each tile, rather than the total image, is then compressed and stored.

Other characteristics of EDMS include:

- the need to cater for regular changes and updates, requiring strict procedures to monitor and control such changes;
- the need to automate the process of modifying old drawings by scanning in existing drawings and cleaning up many of the imperfections;
- the ability to convert scanned manual drawings and convert raster to vector images for subsequent editing in a CAD system. This is an expensive option which is not often used;
- providing a bridge between CAD/CAM systems and existing manual drawing systems.

6.2.3 Workflow/Transaction Systems

A common feature of such systems is the need to capture high volumes of document images as they arrive in a department and to automate the processing of such images in both the front office and also the back office filing activities. Documents are typically handled as items and from the moment the original document is scanned the management of the item is under the control of the application and is routed to different user terminals for specific functions to be carried out. Often there is a link to existing data processing systems to provide the necessary input at the various processing stages. Typical areas which use DIP in this manner are the processing of mortgage applications and insurance claims.

6.2.4 Item Processing Systems

An application which uses a subsystem of the workflow/transaction process with specific requirements is the processing of cheques or credit card vouchers. These areas again require integration with mainstream data processing systems but the main characteristics are the small size of the documents and the very high scanner throughput required when such items are processed. Typically, in clearing banks the number of items processed in a week can run into millions and it has been found that DIP can increase processing throughput by 20–30 percent over previous systems.

6.2.5 Office Systems

The main characteristics of such systems is the ability to capture incoming documents such as customer correspondence, pension documents, personnel information, project documents, such as those required for BS5750, and to manage these alongside documents created on internal facilities, such as desktop publishing and word processing. This can result in the need to manage multimedia documents on one system with compound documents and text documents sitting alongside bit-map page images. The conversion of the textual contents of images from a raster image into coded characters using OCR and ICR and the subsequent integration using Free-Text Retrieval (FTR) software is also a possible feature.

BT use such a system for dealing with customer enquiries within their mobile telecommunications division and recently the Severn Trent Water Authority have installed an IBM ImagePlus system for a similar application.

6.2.6 Mapping Systems

Also known as Geographic Information System or GIS, such systems are used by government bodies and utilities, to provide image mapping of buildings, areas and services.

6.2.7 Publishing Systems

DIP is playing an increasingly important role in the area of electronic publishing. Use of compound document standards such as ODA, SGML and the interchange standard ODIF, have a particular relevance in such applications as publishers start to bring images into their database which have traditionally been text only. The provision of electronic manuals such as the distribution of personnel data in an electronic form rather than hard copy or a procedures manual are further examples of where image has been integrated with publishing systems.

6.3 Case Study Material

In this chapter so far the intention has been to explain the type of application areas in which DIP is being used. In this second half, descriptions will be given of a number of DIP installations from a range of application areas and using different hardware platforms. This case study material has been provided by suppliers and it is a small selection of the many case studies to be found in the *NCC DIP Applications Handbook*. The objective of the handbook is to break down some of the mysteries surrounding DIP and to make it clear that DIP is not limited only to large and wealthy organisations. The provision of case study material from a range of business sectors, covering different applications and using different processing systems, was one way of achieving this objective.

Case Study 1

Subject/company: an Engineering Company

Background and Overview

The user installed Advent's Drawing & Document Management System in order to manage 85,000 of its engineering drawings.

The Problem

The user handled enormous quantities of engineering drawings. These were received from contractors and customers and often required just a simple change, such as a new heading to replace the contractor's heading. A large amount of time was spent redrafting engineering drawings from scratch in order to incorporate these minor changes.

When looking for a system, the user defined the following requirements:

- the system must replace the manual 'cut and paste' operations;
- the system must produce good quality masters from poor originals;
- the system must be able to handle 10,000 drawings in the first year of operation;
- in total, the system must handle 85,000 drawings;
- payback must be made in less than two years.

The DIP Solution

Advent Systems provided a system that allowed engineering drawings to be scanned and then raster edited. Thus, changes were made very quickly. For example, an A0 drawing that simply needs a new heading can be scanned, edited and plotted in under 10 minutes. The system also provides Electronic Drawing and Document Management ensuring that drawing office procedures are adhered to. If necessary, vector data is overlaid and the completed drawing can then be plotted with the overlays.

The system comprised an A0 scanner, A0 plotter and three SUN work-stations and was installed in three months. Cost comparisons showed savings of £94K in this time, based on scanning 994 drawings.

Supplier Information: Advent Systems.

Case Study 2

Subject/company: the Guiness Trial

Background and Overview

The trial arose from the 1986 takeover of Distillers by Guiness whose contemporary share price had a crucial bearing on the value of their bid. The Guiness Chairman at that time, Ernest Saunders, and his co-defendants, Gerald Ronson, Anthony Parnes and Sir Jack Lyons were accused, among other charges, of conspiring to support the share price illegally.

The Problem

Widely described as probably the most complex fraud trial this century, the case paperwork occupied several filing cabinets' worth of seized documents, DTI transcripts and interview material.

The DIP Solution

Unbeknown to most of the participants, the Prosecution had in their armoury an electronic equivalent of this mass of paperwork; a Document Image Processing system called *Catchall* capable of storing the entire case on a single optical disk. Furthermore, as each day's proceedings unfolded, the Court transcript would be added successively to the optical system to provide a complete up-to-date 'bible' of events.

The system chosen by the Serious Fraud Office (as originators of the prosecution) incorporated Intelligent Character Recognition and Free-Text Retrieval besides Image Processing. The major benefit of this approach is that every significant word in a document is indexed automatically; the content of the paper files is therefore searchable at speed; retrieving the 'hits' (such as 3,000 references in 100 documents) would typically take under two seconds; displaying any document, a further three seconds.

Catchall is the brainchild of Surrey-based systems integrators, CACL. Taking, where possible, existing products and modifying them where necessary to produce a coherent whole, CACL is believed to be the first company in the UK to integrate image with free-text techniques. Thus, the system chosen was the Corporate Retriever package developed in Australia by QCOM Pty Ltd. This package has two significant advantages over equivalent products: it is extremely fast in both indexing and retrieval and has a very low index-overhead – crucial when dealing with optical disk capacities.

The Catchall implementation was designed to make retrieval of text and image syntactically identical, making the system easy to use and encouraging the end user to interrogate the database.

Complex searches that would be virtually impossible to undertake with paper files are achievable within a few seconds. The sophistication of the search engine allows for 'wild cards', proximity searching and 'sounds-like' or 'fuzzy matching' techniques. That said, there is naturally a degree of interpretation involved in certain circumstances: the enquiry "Did Jones and Smith meet on 1st May?" could generate several searches on the database to cover possible textual variations such as 'met', 'meeting' and so on. Also, terms frequently used (such as 'Guiness') would more sensibly be combined with other words or phrases to refine the search process. As the Guiness Trial progressed the system was increasingly being interrogated to check precedents in previous days' proceedings, causing much shuffling of paper by the Judge, Defence Counsel and the Jury.

By the end of the trial some 20 Mbytes of transcript material had been added to the original 80 Mbytes of case documents. Closing the case 'electronically' involved the simple transfer of search software and free-text index (a further 20 Mbytes) to the optical disk. A single 5.25" disk now holds the entire paper contents of the 30 drawer filing cabinets and is instantly accessible.

Surprisingly, the total price of the system was under £15,000 – much less than the cost of a single day's proceedings!

Catchall Configuration

286 PC; 640 KB
Kurzwell DISCOVER Model 30 ICR
200MB Optical Disk Drive
High-Resolution Terminal
Free-Text Retrieval Software
Catchall Document Management Software
Laser Printer

Since their initial purchase in March 1989, the Serious Fraud Office have steadily increased their investment in Catchall systems: currently two complete Catchall configurations are acting as input stations, feeding a further six 'basic' configurations consisting of optical disk drive and software. It is anticipated that most of the major forthcoming fraud trials prosecuted by the SFO will involve Catchall.

Supplier Information: CACL.

Case Study 3

Subject/company: a Financial Sector Company

The DIP Solution

A system pilot is to establish feasibility of combining multiple-page legal contract files with loan details and conditions, using an existing accounting system based on a minicomputer with the PICK operation system linked to a DOS/Windows DIP PC network.

The application is for lawyers and legal departments in financial organisations. The application monitors the various securities and legal charges against loans for large capital projects. A single loan can be backed up by numerous securities on assets, property and shares, each of which will often be a separate legal contract, sometimes over a hundred pages in length. Pages of the legal details of the security will be displayed alongside the ledger accounting information on a high-resolution split screen. Any number of pages of contracts can be viewed or printed at will from the workstations in the network.

The application eliminates the need for multiple photocopies of information and stops any change or loss, through misfiling or other causes. View and print stations can be sited in remote locations over communication lines, thus widening the benefits to users in different offices throughout the country.

The system gives the user the best of both worlds, a multiuser system for applications coupled with a high capability DIP system.

The standalone system costs approximately £35,000 and a networked system from £45,000 plus the appropriate PICK server.

Computer Imaging are a London based company supplying total solutions and consultancy for Document Image Processing (DIP) systems for the Financial Legal and PICK Markets.

Supplier Information: Computer Imaging Ltd.

Case Study 4

Subject/company: Warner Chappell Music Ltd

Background and Overview

Warner Chappell Music is one of the largest music publishers in the world with its headquarters in London's West End.

The Problem

The company headquarters was overflowing with important paper documents and needed a way of reducing the filing burden. The Company would remove the bulky files to another site, thereby obtaining better access to their contents.

In particular, the storage of contracts with clients such as Madonna, Prince, and Michael Jackson, was identified as an area of major potential for such a system. These contracts can be anything up to 50 pages long and although many are referred to only occasionally, they are likely to be needed in a hurry.

The DIP Solution

After looking at a number of alternatives, Warner chose an electronic filing system, the EDS Jupiter, from Aldershot-based Electronic Document Systems Ltd. As specialists in the field of Document Image Processing, EDS discussed Warner Chappell's requirements in detail and set up a sample database to scan and index the contracts.

EDS has installed a fast 386 microcomputer, an A3 image scanner, a laser printer, and a very high resolution monitor, together with the Jupiter software and a system that compresses the image files to reduce the storage space. The scanned contract images are stored on an optical disk system that holds up to 15,000 A4 pages on a single 5¼" inch disk cartridge. This equates to a paper filing space of around two or three standard filing cabinets, giving Warner Chappell the ability to re-locate its paper files outside the expensive West End area and utilise the space for more profitable purposes.

Peter Wilson of Warner Chappell's PC Support department says that they looked at the alternatives for about a year before finally selecting the EDS system. "The capabilities of the systems we were shown were all much the same, but what stood out about the EDS package was the simplicity of the image database. Many other systems required us to write our own database software which we didn't particularly want to do".

The EDS system enables Warner Chappell to scan complete contracts and display them on the screen for indexing. This indexing is on the basis of a pre-defined 'template' with common fields that apply to all contracts. The template was set up specifically for the contracts database, but it is simpler for Warner Chappell to set up its own multiple databases and templates according to its requirements. Once indexed, the document images are compressed and stored on the optical disk WORM drive that Warner Chappell supplied themselves.

Once stored, the software enables the users to retrieve and display any contract within a few seconds, simply by entering data into any one of the fields in the template. In this way, a contract can be found by entering the date it was signed, the contract partner's name, the song title or any other index fields. It even allows complex search expressions such as 'all contracts signed in March 1990 with song titles that include the word 'Tomorrow' to be entered for even greater flexibility. With a contract displayed on the monitor, the user can zoom in to an area for greater clarity and even append notes as a reminder. The contract, either in whole or part, can be printed out on the high speed laser printer if required.

Peter Wilson admits that, although they have only had the systems for a short time, it looks like it will be of real benefit. "I think the system is going to be perfect for our needs", he says. "We are also looking seriously at storing our very large files of music on the system. At present these are kept in envelope files which have to be sorted through each time a request for a particular manuscript is received".

By storing the music electronically, it could also help to conserve this extensive library of music, which includes quite rare items.

Supplier Information: Electronic Document Systems Ltd.

Case Study 5

Subject/company: a Registrar's Office

Background

The customer is a government agency responsible for the registration of births, deaths, marriages, divorces and name changes. The agency handles half a million requests for service yearly.

The Problem

Over 18 million registrations going back over 120 years were held, approximately 40,000 volumes of books, weighing about 30 pounds each. The archive was growing, and old documents were deteriorating with constant handling. It was becoming increasingly difficult to handle the large amounts of information.

Customers would make requests for copies of certificates either by mail or in person at a walk-in counter. It used to take many days – sometimes weeks – to fulfil requests.

When a request for a certified copy of a certificate was made, it was entered into an IBM mainframe. An overnight match was attempted against the database of registrations. If a match was found, a report was produced so that staff could go to the file room, pull the document, and make the copy.

The situation was further aggravated because the agency was sited in two locations, over 1,000 miles apart. It was difficult and costly to access the paper-based library from these different locations.

The DIP Solution

Of the 18 million documents, it was decided that 10.5 million should be converted to images. They were scanned at a rate of 65 to 70 thousand documents per day, over an eight month period. A total of 84 temporary personnel working two shifts performed the scanning.

The scanned images are now stored on 250, 12" optical platters. Requests can be actioned immediately regardless of the location at which a request is made. Document retrieval staff each have a PC workstation from which they can make the search and retrieve the associated document image. Existing registration numbers on the IBM system are linked to the images in the Hewlett Packard (HP) Advanced Image Management

System database, and displayed on a single screen. The image is then printed for the customer on a high quality HP Laserjet laser printer.

Introduction of the HP Advanced Image Management System has had direct and indirect payback. The initial study, which looked at the workflow processes involved, highlighted some inefficiencies. The document image management solution now utilises a re-designed workflow. Many of the steps in the original manual process were eliminated or streamlined. The need for a large paper library and its associated paper handling process was also eliminated.

All paper moving, sorting and work division steps have been automated. Imaging has eliminated the cumbersome and time consuming set of steps involved in retrieving and copying the registrations. It has also eliminated the need to microfilm the paper records.

Projected productivity improvements include payroll savings of about 15 percent annually, due to the relative efficiency of the new workflow, based on the image system. Annual savings of 26 percent non-salary operating budget are due to savings in floor space, data entry equipment, book binding and restoration costs, as well as mail and courier costs.

Supplier Information: Hewlett Packard.

Case Study 6

Subject/company: Severn Trent Water Authority

Background

Severn Trent receive a large volume of correspondence, including some complaints, every day both by letter and telephone. Correspondence may refer to water supply discoloration, water leaks, change of address, requests for Direct Debit, etc.

The Problem

They receive on average about 10,000 items of correspondence and telephone calls daily in their Customer Accounting department. For telephone phone calls, the telephonists will write down the enquiry on a form. The letters from customers and the telephone calls (transferred to forms) are then microfilmed so that everyone can have access to every document. The paperwork (letters and forms) is sent to various departments to process. About 50 percent of these work items arrive via the telephone.

Peak loads are 20,000 microfilmed items per day.

The DIP Solution

The Application

Severn Trent have combined the facilities of ImagePlus with Office-Vision.

With the telephone calls, they have used the document format facility of OfficeVision to create different forms for the type of enquiry, whether name and address change, Direct Debit, etc. They have linked these forms to the name and address database from their computer system and the telephone messages are then routed to the ImagePlus queues. This has been proved to be much more efficient than the telephonists completing the forms manually, and then scanning and indexing them into ImagePlus.

The letters are scanned into ImagePlus and routed to the appropriate departments. The departments therefore see the following:

- the paper documents as before but they are in 'image form';
- clearly typed OfficeVision documents (taken from the telephone) that are in the ImagePlus queues. Before OfficeVision, these forms were handwritten, difficult to read, etc.

User Reaction and Benefits

The full scanning volumes, and linkage of OfficeVision and ImagePlus, have been in production for some time now. Routing was scheduled to go 'live' in the third quarter of 1991.

The users from Customer Accounting are delighted with progress to date. In particular, productivity of the Telephone Bank has increased, and customer service has improved because customer phone calls are being answered more quickly. The quality of documented work items has increased greatly as a result of the use of OfficeVision, which has pleased the processing departments.

Customer Accounting are looking forward to the implementation of the next stage – routing of image directly to their desks. The benefits of more up-to-date customer correspondence history files, and better retrieval quality than could be obtained with microfilm, are already apparent.

Timescale and IBM Involvement

There was extensive IBM involvement.

David Garcia (Applications Management Consultant) provided the methodology for a study, which was completed on 2nd November, 1990. The study took 21 days.

The study team was driven very much from the business line. Apart from David Garcia there was a Project Manager (Accountant), analysts from the user area and from DP, an IBM Image Consultant and a Consultant from the accounts team.

The study covered the following areas:

- problem definition;
- requirements definition;
- business process analysis;
- benefits analysis;
- high-level design;

- officeVision/imageplus interfaces;
- detailed design;
- project plan, including resourcing.

The OfficeVision application was available on 4th February, 1991. The letters were printed off and sent to the departments along with the customer correspondence.

Scanning and Indexing went live on 4th March, 1991. Currently, alongside the OfficeVision side, they are now scanning and indexing close to 10,000 documents daily.

They have installed 100 workstations of which 12 are used for scanning.

The implementation plan is now moving ahead of schedule in some areas. For example, the Severn Trent Water Authority are planning to bring forward the use of OfficeVision for outgoing correspondence, which will incorporate those letters into the ImagePlus folders.

Supplier Information: IBM.

Case Study 7

Subject/company: Danish Post Giro

Background

The Danish Post Giro is equivalent to the UK Girobank company.

The Problem

Post Giro needed a faster and more efficient means of signature verification than the traditional card-based method in use.

The DIP Solution

The system provides for the storage and retrieval of some 650,000 sample signature cards relating to accounts held by the Post Giro. The permanent system comprises four A3/A4 document scanners, 16 3.2 Gbyte optical disk drives and 30 Hewlett Packard/Apollo Unix workstations. Additional scanners were provided on a temporary basis during the initial scanning of the same signature cards.

A number of discrete optical disk drives were utilised rather than a jukebox system, since the response time for retrieval of any of the signature cards has to be within 10 seconds. This level of performance could not be guaranteed with a jukebox system. All of the workstations can be used for all functions in the system, ie there is no requirement for dedicated scanning or viewing stations. The function of any particular workstation is dependent on the job function of the operator using it at any particular time. Each workstation also acts as a terminal to the bank's IBM mainframe system. Thus, a query to the mainframe database can result in the desired signature card being displayed on the workstation.

The system was developed and installed by GN Filetech using their ODIN document image management software package. In addition to the installation at Danish Post Giro, the system has now been adopted by both the Belgian and Swedish Giro banks.

Approximate cost of the Danish installation is around £500,000 including full application customisation.

Supplier Information: Iris Solutions.

Case Study 8

Subject/company: Royal Life Holdings

Background to the Company

Royal Life Holdings in Liverpool is one of the beneficiaries of the explosion of new activity in the pension and life assurance business. A wholly-owned subsidiary of Royal Insurance Holdings plc, Royal Life has wide ranging interests in financial and service sectors from estate agency to executive pensions and has enjoyed massive growth – there are now over 10,000 staff in the UK with many thousands more worldwide.

Royal Life were early to recognise that imaging technology offered the best solution to its paper handling and customer service problems, particularly within the Executive Pensions department.

The Problem

More than two and a half million sheets of paper pertaining to 'live' executive pension plans were stored in the company's basement. The problem of volume was further compounded by the complexity of the business.

All Royal Life's archiving is on-site. This did not pose immediate problems in the relatively affordable Liverpool area, but it did mean that as business accelerated, Royal Life was having to acquire more and more filing space. In addition, market pressures demanded shorter time to market for Royal Life's pension products.

The company's IT department and pensions management analysed the benefits of DIP and concluded that their own specific needs could best be met by using the Olivetti FileNet system to deliver both mainframe and paper filed information direct to the Pensions staff by means of a series of task driven routines.

The DIP Solution

Royal Life Project Manager, David Cowan, commented that"Olivetti's FileNet offering was the only system that could fulfil all the specifications required. Price coupled with functionality made Olivetti our choice".

The initial configuration consisted of a dual server system with as

64-slot OSAR (Optical Storage And Retrieval library), Two scanners, 15 clustered workstations and a printer; subsequently increased to 24 screens and an extra printer. The total investment for the system has been approximately £750,000.

A key requirement was the ability to integrate the image processing system with IBM operating systems. Small expert systems have been implemented, enabling the FileNet system to pull data from the mainframe and merge data with images.

The system is expected to substantially improve the company's response rate to correspondence and queries from a previous wait of as long as 10 days to just 24 hours. Service is improved because the collection, processing and analysis of information has been automated.

At Royal Life, WorkFlow eliminated the information 'float'. It specified precisely what information to retrieve and when and where to distribute it. WorkFlo routes the information as well as orchestrating and controlling the unique processing tasks to be performed at each workstation.

The key to successful implementation of a DIP system, is to start from a comprehensive understanding of a company's business requirements. Royal Life's installation is a complex application, where a number of background activities take place 'behind' the screen, enabling tremendous increases in processing times to take place.

This advancement has only been after in-depth analysis by both Royal Life and Olivetti, followed by expert development of the system and the WorkFlow software. David Cowan concludes "Our understanding of the system and its capabilities point to considerable scope and/or expansion".

Supplier Information: Olivetti Systems and Networks Ltd.

7

Types of Systems

7.1 Introduction

Like many of today's computer applications DIP can be offered on a wide range of computer platforms just as text or data can be created on almost any system and the case studies described earlier certainly bear this out. This chapter looks at these different platforms in a little more detail. What has to be considered is that, in the longterm, DIP should not be seen as an island technology and the major returns on investment are often where DIP is used in an integrated environment and not as a standalone system.

The saying that there are many ways to slice a cake is very applicable when it comes to describing the different types of DIP systems. In one sense they can be classified according to the various applications described in Chapter 6 and many vendors do in fact use this classification. However, it is also possible to break the DIP market into large, medium and small systems which equate to the mainframe, mini and Personal Computer of the general computing arena as well as describing them in terms of level of integration.

7.2 Classification by Size

7.1.1 Large Systems

Such systems are based on medium to large CPUs, such as offered by an IBM 3090 series, and as such are very expensive. They are intended for company-wide applications and can support hundreds or even thousands of workstations. With the present state of DIP usage, in the UK there are not that many systems of this size yet installed although they are more common in the USA.

It is predicted that as confidence in the use of DIP grows, most major mainframe vendors will announce products in this category as part of their overall imaging strategy.

7.2.2 Medium Sized Systems

Such systems tend to be used to handle smaller departmental applications or even those applicable to smaller enterprises. Small- to medium-sized CPUs such as the MicroVax or AS/400 are common in this arena and the number of workstations supported can be as high as 200 or 300 and as low as 8 to 10. The Filenet product fits into such a category, as does the offering from Wang.

7.2.3 Small Systems

This classification is based on the use of microcomputers and can be divided into:

- standalone systems;
- networked systems.

Standalone PC systems

An 80286 or 80386 based microcomputer together with a scanner and optical disk makes up a basic DIP system and, as is highlighted in the NCC DIP Applications Handbook there are an increasing number of systems suppliers providing solutions to the marketplace and also applications for which this PC-based approach is being put foward. They are intended for small, work group applications or single users and are being purchased to give companies an insight into the use of DIP.

Networked PC systems

Such systems use the same basic 286 or 386 based PC linked together over a Local Area Network.

7.3 Classification as Centralised or Distributed Systems

The traditional centralised approach to processing systems has changed and distributed processing is now the favoured approach. This has been as a result of powerful processors being available at lower costs making it possible to have processing power available as a local resource, which can be dedicated to a set of specific tasks such as those associated with DIP. Because DIP requires large amounts of processing power it would not be possible to support multiple DIP users on dumb terminals linked to a central host computer

7.3.1 Centralised Systems

The first DIP installations tended to be standalone centralised systems based on PCs. It can be said that the single user PC systems, which are entering the market in increasing numbers, are centralised systems in which the DIP peripherals such as the scanner, are interfaced to the PC by specialist boards and are controlled directly by the central micro-computer. Because of the limitations of centralised single-user PC systems they are only used in relatively low volume applications, or as input stations when users expect in the longer term to expand into a networked multi-user system.

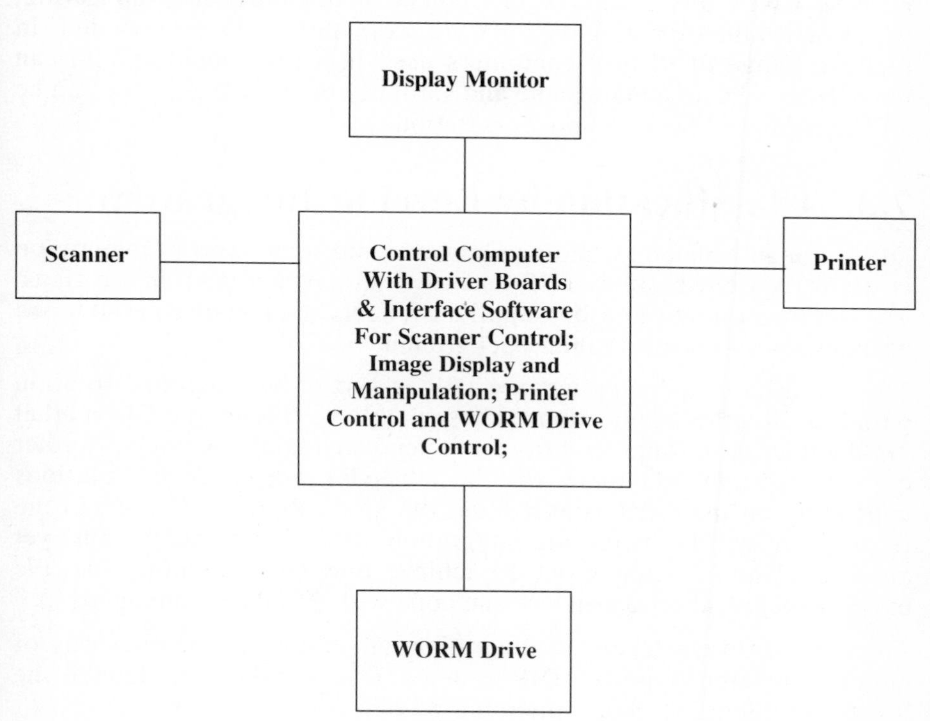

Figure 7.1 A DIP system based on the centralised processing approach

7.3.2 Distributed Systems

A distributed processing approach is the preferred option for most multi-user DIP systems. Such a system would use distributed processing and consist of multiple PCs, or workstations linked to a central file server. A typical configuration for a low-volume applications would be:

- a 386 PC acting as a file server with magnetic and optical storage;
- a 386 PC acting as a DIP server with interfaces to scanning and output devices;
- several standard PCs acting as viewing terminals;

- such terminals linked to the file server and the DIP server via a Local Area Network. An example of such a configuration is given in Figure 7.2.

In larger scale applications the PCs have been replaced in certain circumstances by more powerful workstations such as those from Sun. The DIP server and file server role being taken on by a large general-purpose UNIX box. There is an element of flexibility now appearing, which will allow existing PC or PS/2 workstations to be included. In fact, the range of control computers used in a distributed system can range from a PC to a mainframe and similarly the viewing device can be a PC, an Apple Mac or a Sun workstation.

7.3 Classification by Level of Integration

DIP is not an isolated technology however and some level of integration is often required between DIP and existing systems and applications. The approach advocated by suppliers reflects their own strengths and weaknesses and possibly their background.

The major hardware suppliers see DIP as part of an integrated solution with links to other applications. Those that have entered the DIP market via the microfilm route tend to recommend standalone systems. Neither extreme is right and there will be a need for a range of solutions depending on the circumstances and the applications. However, in the future there will be increasing integration, as DIP must not become yet another island of automation. To achieve integration assumes that the existing technical infrastructure can cope with this development.

There are different levels of integration which give yet another way of defining different types of DIP systems. One supplier has defined the levels of integration in the following ways.

7.4.1 Standalone Systems

This first classification in fact provides no integration at all. The DIP system is implemented as a standalone system with no external communications and acts as an electronic filing cabinet. As mentioned earlier, this may also be an approach that some companies may take as an entry point into DIP with the option of adding communications links at a future date. This can be a risky option if the suppliers progress towards an integrated system is not fully evaluated.

7.4.2 Terminal Emulation

This second level of integration is basically a standalone DIP system in which the DIP terminal is also physically linked to another host computer

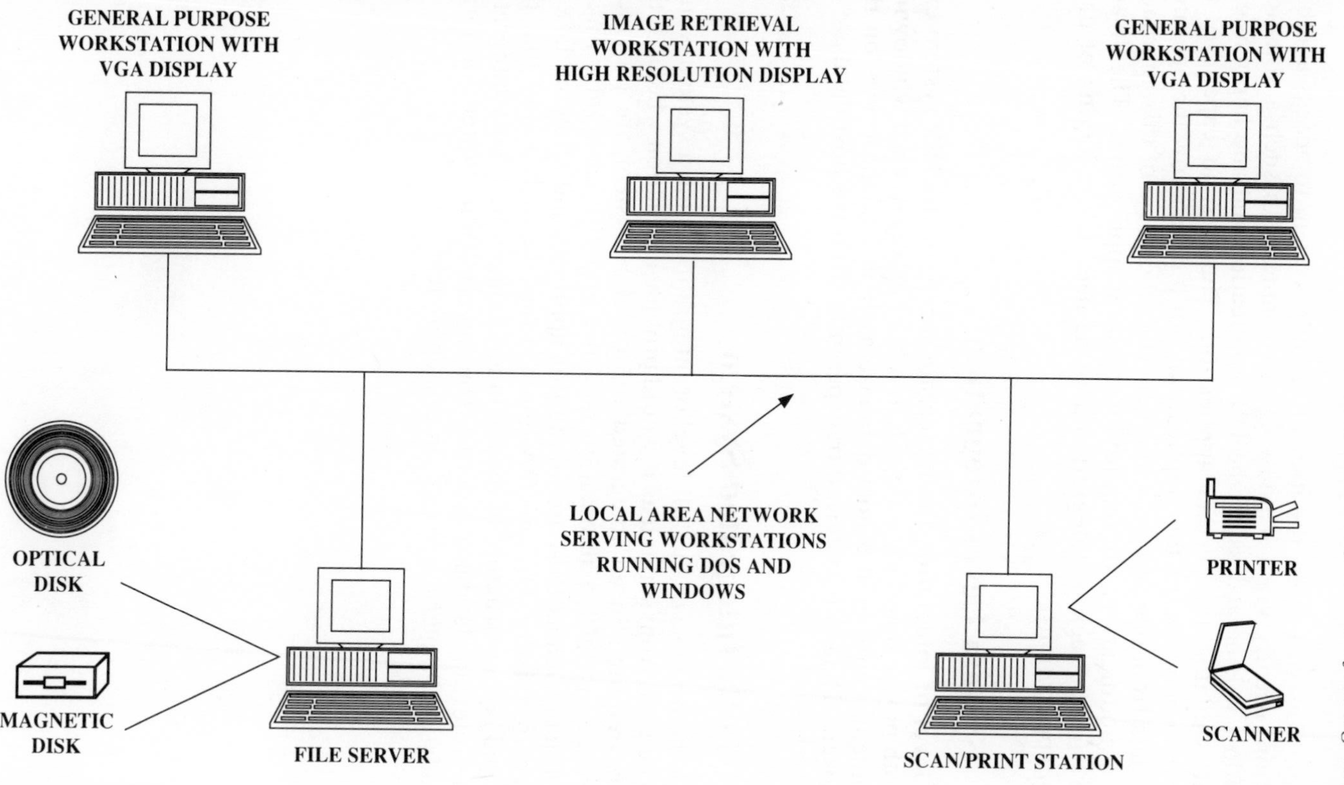

Figure 7.2 A generic PC-based distributed DIP system

computer. This will allow the user of the DIP system to access another application, usually via a windowing environment. In such a situation one window will be accessing the DIP application and another, using a terminal emulator, the second application. There is no capability to transfer data between the two applications but it does allow the operator to enter data from one screen to another. The advantage of this approach is that no modifications are required to existing applications. The disadvantages are the lack of integration and the need for re-keying of data between applications.

7.4.3 System Level Integration

At this level of integration the image/file server of the DIP system can make both physical and logical links to other hosts and, via a windowing environment, can search a host database as if the database was on the local system. Data is transferred transparently from the host database to the image server.

7.4.4 Fully Integrated System

To make the best use of the full level of integration, such a system would have to be built from scratch as it would provide a situation where all the functions required in an integrated information system are provided. Data processing, office automation facilities and DIP are tightly linked and suppliers such as IBM are taking this approach and allowing DIP to be added to existing DP and OA applications, in such a way that the logical and layout structure of the document images are totally integrated with the relevant database records. The Severn Trent Water Authority case study in Chapter 6 is based on this approach.

8

The Technologies of DIP

8.1 Introduction

The vast range of applications to which DIP has been utilised means that no single description can be given of the imaging process, as each different application may have different requirements. However, regardless of the application the various stages of the imaging process can usually be broken down into six basic steps. These are:

- *image capture* – documents to be scanned are fed through a scanner which converts them into a digital form;
- *indexing* – all images have to be indexed to be able to retrieve them from storage;
- *storage* – the electronic images are usually stored on an optical disk but they can be stored on any magnetic media;
- *retrieval* – the images are retrieved and displayed on a workstation;
- *manipulation* – images can be annotated or even edited;
- *output* – images can be output to printers either in a local or remote location.

In the first part of this chapter the processes used to perform these basic steps will be examined and a description given of workflow software which often supports the technology of DIP.

8.2 The Processes of DIP

The processes which collectively make up a typical DIP system are outlined below and Figure 8.1 provides an overview of the six stages.

8.2.1 Image Capture

The first step in the imaging process is to convert the document to be captured from a hard copy form into a digital form using a scanner.

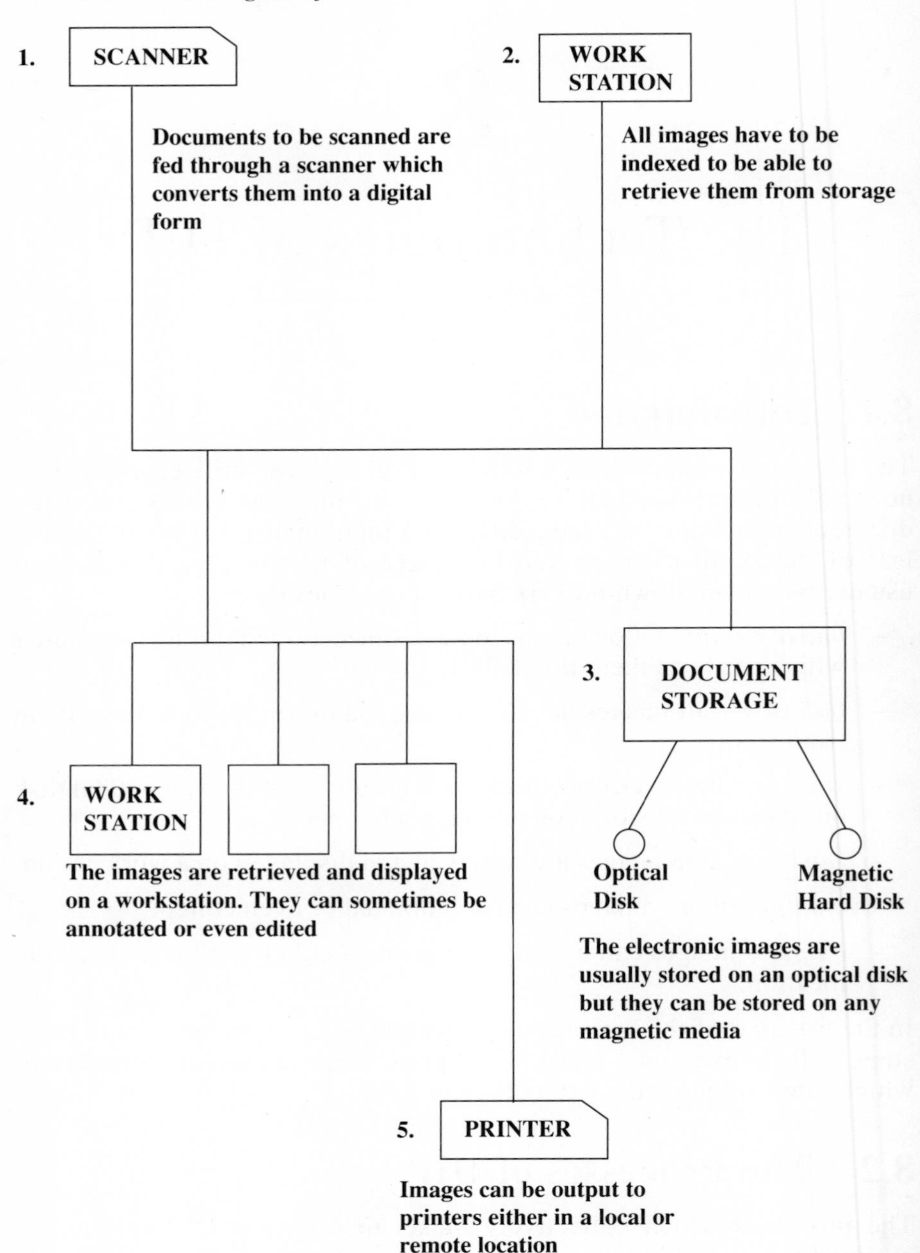

Figure 8.1 Overview of the document imaging process

Such a scanner works in much the same way as an office copier and scanners of one sort or another have been in existence since the 1920s.

A scanner passes a source of light across the document and by measuring the intensity of light reflected by the document can recognise areas of dark and light. The scanner then creates a bit map or raster image of the document.

The file containing the bit map image also includes additional data that defines the resolution of the image, its size and other characteristics such as information needed to be able to view the image on a workstation or print out a hard copy.

At this stage the computer sees the image only as a series of '1's and '0's corresponding to the light and dark of the original document. Human intervention is usually needed at this stage and part of the image capture process relates to naming the image file and checking for quality. Once the document has been scanned the image can be transferred to a workstation where it can be allocated a name and checked. At this stage the image is totally unintelligible to a computer, as the scanned image is held as a raster file and therefore it is important to allocate to it a unique name to be used to subsequently retrieve the image. Naming of each individual file by an operator is only practical in small scale installations. In high-volume systems using automatic document feeders, filenames are normally assigned automatically.

An additional process before storing the image is to check for quality. In certain circumstances the scanner may need adjustment and the documents may need to be re-scanned. It is also at this validation stage that a system operator may have the facility to edit the copy of the original document. The types of typical facilities available on image editing packages include:

- speckle removal;
- crop, cut and paste;
- copy facilities;
- zoom and size of scale facilities;
- erase.

The level of image enhancement facilities provided will depend on the DIP application, with high-volume applications using minimal facilities compared to the wider use in office systems.

Increasingly Optical Character Recognition (OCR) capabilities are included in DIP systems as a means of recognising alphanumeric characters and translating them into an ASCII format that the computer understands.

8.2.2 Indexing the Image File

Once the image quality is acceptable the very important task of indexing the file takes place. The operation makes use of a document database management system which typically uses a customisable menu to include reference fields for all the information needed to properly index the image.

Such a system would include the following fields;

— name of the image;

— date of scanning;

— size;

— relationship with other images;

etc.

Automatic indexing using OCR facilities to read-in information from the original document is also a possibility and is used in high-volume applications.

The subject of document databases is a major issue for any DIP installation and is dealt with in more detail later in this chapter.

8.2.3 Storing the Image

Once the indexing processes has been completed, the permanent storage of the image can take place. The image file together with the indexing information is sent to a document storage processor via an appropriate network, linking all the components of the DIP system together. It is the role of the document storage processor to handle the storage of the image and its associated indexing information.

They are not stored together as normally the indexing data is stored in a database structure held on a magnetic disk. The image meanwhile can be stored on either a magnetic or optical disk. Most DIP systems today make use of optical disk and the most common form of optical disk storage used in document imaging systems is a WORM (write once, read many times) drive. This form of optical storage is permanent and cannot be overwritten.

For audit purposes DIP systems may also make use of a journal facility to record all storage and retrieval activity. There may also be a magnetic disk cache used to store recently retrieved images. Use of this facility can reduce the time taken to retrieve images as the access time for magnetic disks is faster than for optical disks.

8.2.4 Retrieving Images

The time spent setting up your document database is justified when you come to retrieve images. The query language of the database is usually a very powerful tool for retrieving either a single image or possibly a number of related images. The more indexing information entered into the database, the more retrieval options are open to the users.

Individual images can be retrieved from their optical storage device using a unique image reference, a creation or modification date, or any other criteria which has been included in the initial indexing process. The images are carried over the network to an appropriate processing workstation.

On the basis that the workstation has the appropriate editing and markup software images once retrieved can be annotated or modified.

8.2.5 Annotating and Modifying Images

Images transferred to a workstation with editing and markup software can be modified or have comments added to them. A typical DIP application which would use such facilities would be EDMS (Engineering Document Management System).

For example, once an engineering drawing has been retrieved to an editing workstation, comments can be added or even proposals to change the design shown on the drawing. However, the original image is not in fact changed but rather the additional information is held in the form of an 'appended' overlay. This is usually a processable vector file which uses a language that specifies x-y coordinates, arcs, etc, as distinct from the original engineering drawing image which is a bit-mapped image based on pixels.

The annotated material can be stored on either magnetic or optical disk. Information relating to the relationship of the annotation to the original image is entered into the document database and when the original image is retrieved, the annotations or modification overlay is automatically appended.

Some changes, however, may be of a permanent nature and although in the engineering example, the proposed alterations will be made on a vector overlay, this can subsequently be converted into a bit-mapped format and incorporated into the original image. The normal process is to store this revised image as a new file without over writing the original and to make appropriate changes to the document database so that an update trail can be provided.

8.2.6 Image Output

Most DIP systems will contain a print or plot server at a workstation so document images can be output in a hard-copy form. Experience has shown however that this is a facility which must be controlled as in the initial stages of implementation, high volumes of printed images are often made. Such facilities are usually available via any workstation on the DIP network and the menu used to retrieve an image usually includes a request to print out selected images.

8.2.7 Workflow Software

An increasing number of DIP systems are now including additional software known as workflow. This provides a facility to automate the white collar clerical/administrative operations in much the same way as blue collar jobs have been automated on the shop floor.

Typically workflow software uses artificial intelligence technology to control the sequence and time that a document moves through an organisation. Having recorded in detail the present procedures and made changes to reflect actual needs rather than historical precedence, a script can be 'written' using the workflow software which will control the flow of a document as it is routed to various people who need to work on it.

A simple example of the use of workflow software is in a DIP system used for processing incoming invoices for example:

- when the invoice is scanned it can be automatically routed and stored in the appropriate invoice folder, which is in turn automatically routed to the first person who should receive it.
- when this person receives the folder, it might also be placed in a job queue monitored by the system. This ensures that if the invoice processing clerk does not work on the invoice within a specified period, a reminder can be issued by the system that this invoice needs actioning.
- when one clerk has completed work on an invoice, such as checking prices and addition, it can be automatically routed to the next appropriate person, who may for example be able to approve the invoice.
- such software can be used to establish work priorities and optimise allocation of work to individuals, as well as tracking documents as they progress through the system.

8.3 The Technologies of DIP

Having briefly examined the processes associated with DIP, it is possible to described the technologies which are used to perform those processes, many of which will be familiar to existing users of office automation facilities. How they have been put together and work as a DIP system is described in the remaining part of this chapter.

This chapter can only give an outline of DIP technology and anyone interested in more details should refer to the *NCC DIP Reference Volumes* and in particular the volume called 'A Technical Introduction'.

In the following pages the technology will be looked at in the logical sequence of the DIP capture and processing cycle, as outlined earlier.

8.4 Scanners

Scanners are used in the first stage of the imaging process to capture images of hard copy documents in an electronic or digital form. They work in a similar way to the traditional office photocopier, using a light source, typically a fluorescent lamp and a light sensor. The light reflected from the document is picked up by light sensitive cells on the light sensor, with each cell receiving reflected light which equates to a single dot or pixel on a line of a page. The intensity of received light is recorded in the form of an analogue signal and is analysed to determine whether the scanned pixel will be interpreted as 'White' or 'Black'. The signal is then converted into a digital form and is stored as a bit map or raster format file.

The chief factor which determines the quality of the scanned image is the resolution of the scanner and the amount of grey scale that the scanner can detect. Scanning resolutions are expressed as a number of dots per inch (dpi) in a horizontal direction across a document and in a vertical direction down a document. A typical resolution for document scanners is 200 dpi which is compatible with the resolution of a Group III fax machine. Higher resolutions of 300 dpi or even 400 dpi may be required for certain applications which include, for example, small font sizes, fine lines or to improve the grey shade capture of photographic material.

There are basically two different designs of scanner – a rotary and a flat-bed scanner. A flat-bed scanner tends to be used to scan bulky items such as bound books or magazines. They can also incorporate automatic sheet feeders which allow anything between 10 to 100 sheets to be stacked and automatically fed into the scanner. Rotary scanners are usually sheet-fed and can also be supplied with an automatic sheet feeder. Where single sheet pages have to be scanned consecutively rotary scanners tend to be the preferred choice. Figure 8.2 outlines the scanning process.

The file that contains the bit map or raster image of the captured document

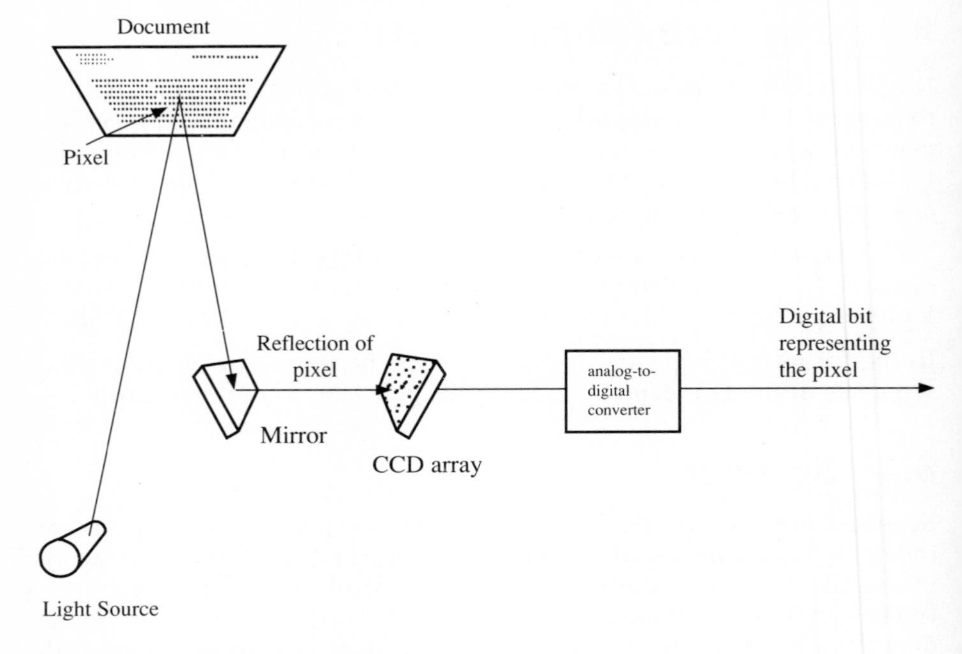

Figure 8.2 The scanning process

also contains additional information which defines several characteristics of the captured image such as its size and resolution. This data is required in order to display the image on a screen or output it to a printer. What is often misunderstood is that although the bit-mapped or raster image can be displayed on a computer screen, the contents of the image are unintelligible to the computer. It only sees a large number of '1's and '0's corresponding to the light and shade of the document. For this reason a distinction needs to be made between the scanners described so far, which are often known as page image scanners, and text scanners.

8.4.1 Use of OCR and ICR

Page image scanners capture an image in the form of a vast number of dots which allow an image to be displayed but which is not understood by the computer. Text scanners often known as OCR (Optical Character Recognition) or ICR (Intelligent Character Recognition) scanners contain additional intelligent or analytical capability which can view, recognise and code each printed text character or symbol contained on a page. The resultant ASCII file can be subsequently manipulated using a variety of applications software, such as word processing, publishing or text retrieval.

Template matching recognition logic compares the scanned bit pattern with a range of different stored character patterns of a type style or font.

Topological analysis is a more complex approach as it works out various topological details of the scanned character (eg, loops, horizontal lines, vertical lines, lines crossing, etc) and recognises a character by comparing these characteristics with those of generic characters (character types) which it has stored. Template matching machines are cheaper but may be limited to reading documents of a known number of limited fonts and where the quality of documents to be scanned is good. The topological or intelligent character recognition approach may be preferred where the fonts and character patterns are less predictable; where the printed characters are less predictable; and where the user application may justify a more sophisticated approach. Increasingly there is a tendency to use the term OCR in a generic sense, covering both OCR and ICR.

Within the context of DIP these recognition facilities are used for different reasons. They are:

- data entry;
- data conversion;
- automatic indexing.

Some transaction processing applications may wish to automate the data process facility and link into the image capture process. A well-known application is in high-volume items or cheque processing where in addition to capturing an image of the cheque, the data contained on it such as the sort code or account number, is captured so that data can be input into relevant DP systems.

Increasingly in office related applications, users are looking to store not only a bit map image of the original document but also a full text version of the document which can be integrated using a text retrieval facility. OCR/ICR is required to convert the image into a machine readable format.

OCR can also be used to provide an automatic indexing capability. By applying OCR capabilities to a predefined area of an image during the scanning process, information held within this area can be used to index the document.

An OCR capability is an option in many of today's DIP products and Chapter 6 contains examples of the use of OCR in DIP installations.

8.4.2 Types of Scanners

There are a range of scanners available which will accommodate documents of different sizes and with different characteristics. They include:

- *bi-level scanners*, which can accommodate a range of document sizes and aperture cards and in which a pixel is represented by 1 bit;

- *grey scale scanners*, which support the scanning of black and white photographs and also continuous tone documents. Such scanners are able to recognise many shades of grey;
- *specialist scanners*, which are able to accommodate high speed scanning as required in cheque processing and also colour scanning.

Scanners can accommodate a variety of media from paper through to aperture cards. They can also read pencil written information through to typed text. The speed of scanning is dependent on a number of issues but typically the average speed for an A4 page of black and white text, scanned at a resolution of 200 dpi is three to six seconds per page.

The price of scanners varies widely and can range from under £2000 for a PC based product to a flat-bed scanner at £5K and a high speed scanner of between £70 - 150 K.

8.4.3 Compressing the Image File

An A4 sized document, scanned at 200 dpi and with one storage bit used per scanned point, produces a bit map of about 3.74 Mbits or 0.4675 MBytes. This compares to 2 – 4KBytes to store a standard page created on a word processor. In order to reduce the demands on processors, networks and storage devices, the image is compressed using different algorithms. Compression algorithms, such as those used for Group III and IV fax, reduce the size of the bit map by a figure of between 5 to 15 : 1. This means that a compressed image may occupy about 40KB of disk storage and approximately 25,000 pages may be stored as digitised images on either side of a 12" optical disk and 5,000 pages on either side of a 5¼" disk.

Once compressed the image is stored in a raster format, typically in the Tagged Image File Format (TIFF). This format is not only used for image capture but also for facsimile and other telecommunications applications. This format was developed by Aldus and Microsoft as a common format for developers of scanners and desktop publishing systems and is used to provide information and parameters relating to the stored image in tag fields.

8.5 Workstations

Workstations are used to view, manipulate and process images. In ideal circumstances, use would be made of the new generation of personal computers or workstations that are 32 bit machines with several MBytes of memory. In addition they would have many MIPS of computing speed, a large high resolution graphic display capability and extensive storage capability, incorporating multi-tasking and windowing facilities that allow several tasks to be perform simultaneously. What must not be forgotten, however, is that in most organisations there has already been extensive investment in everything from PCs, PS/2s and Apple Macs, through to more- or less-powerful workstations and a vast mixture of terminals. Whether use should be made of existing equipment or not will depend on the application and also on how much effort the user organisation is prepared to put in and the type of equipment in use.

In respect of the ubiquitous PC, what must not be forgotten is that it is almost a generic title under which fits a large range of equipment which is used for a diverse range of applications. In the main PCs are used for one application only and although they are often linked into a network to share common printers, etc, users don't want to be hooked in to every other application. When considering the possible use of existing PCs as image workstations therefore, one must take into account such issues as basic memory availability, hard disk storage capacity, the number and type of expansion slots available. Suppliers of image add-on capabilities assume that an 80286 or 80386 machine is at least being used although even these can have wide differences.

It is possible to use a PC but the PS/2 is more suitable and many board suppliers have developed products for the PS/2 and a number of systems integrators use PS/2s as DIP file servers and workstations.

A range of more powerful UNIX-based workstation are also being used and with their high processing power, advanced graphics capability and operating systems they are ideal for many DIP applications, especially at a departmental level.

The Apple range of microcomputers are gaining increasing acceptance as the standard company workstation but as yet there are few suppliers who have built DIP capabilities into them. However, the number is growing and there are examples of very successful DIP systems based on the Apple Macintosh Plus, SE and II models.

DIP workstations may be required to support a range of activities. These include:

- controlling the scanning process as outlined earlier. It is via a workstation controlling the scanning process that scanning parameters are set, filenames are assigned and the quality and size of the scanned images are controlled;
- storing and retrieving documents by adding indexing information and subsequently retrieving stored images based on that indexing information;
- displaying all or part of a document which may include the use of a range of facilities such as rotation, magnification and overlays;
- adding comments to images through the use of annotation and

modify facilities and overlays without affecting the original image file;

- editing images or carrying out image enhancements using the techniques described earlier;
- controlling printers and plotters;
- providing specialist facilities such as mirroring, drawing capabilities and image manipulation.

To support such activities the ideal workstation used in a DIP configuration should have a high resolution colour monitor, typically capable of a 200 dpi resolution which equates to displaying 2200 horizontal scan lines with 1720 pixels per line.

8.6 Document Database Management Systems

The true brain of any DIP system lies within the document database management system which is made up of two components; the database management software which controls the underlying data storage structure and provides the user interface; plus the storage processor to which are connected any optical or magnetic drives.

It is through this part of a DIP system that all images are stored and retrieved. It also controls access to information and provides basic administration facilities for the system.

It is not the intention in this brief introduction to talk in any detail about hardware issues other than to make some general statements. The computer that runs the database software normally includes magnetic disk storage capability for storing the database files and also any frequently accessed images. The reason why magnetic media is used relates to frequency of access and the resultant requirement for speed. Database files have to be frequently accessed and modified as new images are brought into the system or existing ones modified.

The workstations that provide the interface to the system for the users and also control the scanning process are linked to this storage processor via a network. The operating system that is selected for this processor should:

- be suited to database operations;
- be well suited to handle disk I/O routines and operations;
- should have a good file structure;
- be well suited to work in a networked environment.

In general purpose systems, DIP will normally be based on a general purpose host running a standard operating system such as DOS, UNIX or one of the major proprietary operating systems. However, it must be said that UNIX is the most popular operating system among start-up developers of multi-user systems and many of the major computer companies are building support for UNIX alongside existing proprietary operating systems.

The document database management system should provide a number of facilities. These are listed below in a logical sequence:

- the facility for an operator to index an image and to add any related information which would be of value and would provide 'intelligence' to the storage structure;
- the automatic handling of the database structure;
- the automatic storage of images and related information;
- the facility for an operator to retrieve either a single stored image or a number of related images with common attributes;
- administrative routines such as access control, audit trails and statistical information.

The current state of database technology provides a wide range of development tools and query languages that give flexibility to system developers, particularly when creating user interfaces and storage structures to meet the different requirements from different customers.

Of the two most popular database structures, hierarchal and relational, it is the latter which is possibly more suited to the storage and retrieval requirements of DIP as indexing information held in different tables for different applications can be linked. Consequently a query on one table would result in information being made available from another table if required.

8.7 Retrieval Options

There are a number of strategies used when searching an image database prior to retrieving an image. The following is a brief list of the most common ones:

- use of common attributes in which one field in every record is examined for a specified attribute, such as a unique customer reference;
- use of a combination of attributes in which several fields are examined to find all records which contain the same value in each specified field;

- use of multiple combinations in which several fields are examined to determine if they contain an acceptable range of values.

It is possible to plan a retrieval strategy in advance, especially when the DIP application is not operating in realtime. Using such a strategy can reduce the amount of time spent looking for a document and can ensure all relevant images are stored locally thus reducing the burden on the communications network.

When operators put a query to the system, they are presented with a list of those records which meet the specified criteria. All of these records can be examined or the list can be further refined by issuing additional queries which limit or expand the search.

8.8 Security and Access

Security measures can be implemented when using a database approach which can provide a range of access facilities. These include:

- 'Read only', an operator can retrieve and view the image but is not authorised to carry out any modification;
- 'Modify', as well as retrieving information and images, new modified versions can also be created;
- 'No access', which as its name implies prevents certain staff from accessing information held in the DIP system.

8.9 Storage

A wide variety of storage devices can be used to store images. The traditional digital storage offerings associated with most computer systems such as fixed magnetic disks, magnetic tape and floppy disks are not always suitable for use within DIP systems. Consequently, optical storage media and drives are now available which offer system suppliers, integrators and users an alternative storage option with many attractive attributes required for the efficient storage of images.

The prime reason for looking for an alternative storage medium is that the data volumes that have to be handled in DIP systems are many times greater than we have been used to in most computer applications. For example, an A4 black and white document scanned at 200 dpi and compressed using CCITT Group IV compression standards, can still occupy 25 to 70KBytes of data. If large sized documents such as engineering drawings have to be captured the storage requirements are considerably more. When you consider the number of documents that might be captured daily in a DIP system it would not be economical to use a fixed magnetic disk for the long term storage of all the documents and there is a need to consider a cheaper alternative, especially if it has a higher storage capacity. However, optical technology is usually considered as a long-term archival medium and magnetic tapes provide the most cost-effective backup solution.

The following figures give an idea of the number of standard pages which can be stored on a 12" optical disk.

Number of Pages	Disk Storage on 12" Disk
3,000 (Filing cabinet drawer)	5 percent of disk
12,000 (4-drawer filing cabinet)	18 percent of disk
600,000 (50 cabinets)	9 disks

Optical disks are made of a rigid, composite material with a photosensitive recording surface. A transparent protective layer covers the surface of the disk and prevents contamination. Within a disk drive, a laser is housed together with a mechanism for locating the laser beam on the appropriate part of the disk. A lens directs light onto the surface of the disk as it spins and detects patterns reflected off the disk's surface. When recording data, the laser makes a series of marks on the disk which correspond to bit-mapped data. Conversely a low intensity laser scans for the presence or absence of such marks when reading the disk.

There are now a range of optical storage media available in the computer world which all use a laser beam to record and subsequently read back the stored data. As well as optical disks there are optical tapes and optical cards but for most purposes, especially bearing in mind the introductory nature of this publication, one needs only to consider optical disk technology and in particular the use of WORM optical disks. This fact has resulted in considerable interest in the technology from traditional records management staff who have for some time been looking for an efficient digital alternative to microfilm and paper for document storage.

WORM disks come in a variety of shapes and sizes as well as formats. There are 5¼" and 8" as well as 12" and 14" but despite the apparent problems that this may cause there are a number of reason why WORM technology is suitable for DIP.

In particular, optical disks provide two primary benefits.

- extensive storage capacity. Use of laser technology enables large numbers of images to be stored compactly;
- long life and durability. Unlike magnetic based storage media, they are not subject to head crashes, they are impervious to climatic conditions and are not affected by magnets.

Other benefits include:

- no need for specialist storage of disks;
- much faster access than tapes.

Listed below are the storage capacities of the most commonly available optical disks

Disk Size	Capacity (GBytes)
5¼"	between 0.2–1
8"	between 1–2.5
12"	between 2–6.5
14"	between 6.8–10

Because of the high storage capacities and non-erasable characteristics, WORM technology is ideal for:

- applications that involve large volumes of data such as DIP systems, or replacements for more traditional storage media such as COM, paper, and magnetic tape;
- applications that involve data that need not and should not be altered and for applications which require a tamper proof audit trail.

Although a major disadvantage is often said to be the lack of standards for this technology work is under way to provide for the interchange of information on removable disks by specifying the format of the recorded structure that contain descriptive information about volumes and files or directories recorded on the media.

8.9.1 The role of magnetic media

Magnetic media still has a number of attributes which make it attractive to use for certain operations within a DIP system.

Magnetic disks

These are normally used in a DIP system for storing data that needs to be readily accessible and they can provide faster access than optical disks. This factor along with the fact that they can be written to or read from, means that they are often used in DIP system as a temporary storage such as when images are being compressed or when an image is being revised. They also proved interim storage after a document has been scanned and for the document management database.

Magnetic tape

This also has a role to play and in particular as an off line archival store.

8.9.2 The Jukebox

The removable nature of optical media has lead to the development of automatic optical disk autochanger devices or jukeboxes, which operate along very similar lines to those devices so well loved in the sixties and seventies. Such devices are capable of holding high numbers of disks which can be automatically retrieved and placed in optical disk drives without the need for any manual intervention. Information stored in this way is said to be on 'near-line' storage which in essence is slower than on-line, but offering a much faster response than traditional off-line facilities. Response time for retrieving documents from a jukebox are between 30 seconds and a minute.

Which although sounding better than the time often taken to retrieve paper based information is still a long while to wait at a workstation and one of the reasons why alternative retrieval strategies have to be considered.

8.10 Printers

Although a wide range of printers can be used in imaging systems, everything from a dot-matrix printer to a plotter, the most commonly used printer is the laser as it offers both speed and high resolution. What should not be forgotten is that one of the objectives of a DIP system is the control of paper-based information. Consequently, print facilities should not be made readily available as with the present culture for paper-based information, staff are all too keen to produce a hard copy of a stored image.

The selected printers are often controlled by software within the imaging system which provides for a range of additional features such as:

- printing only a part of an image as well as all of it;
- printing images at a range of resolutions;
- producing scaled down images or cropped images for a particular output device.

8.11 Networks and Imaging Systems

In considering the network implications of DIP, the fundamental issue which must be taken into account is that the size of image files are very large and will have considerable impact on any network. Typically an A4 page even in a compressed form will be 25 times larger if stored as an image rather than as an ASCII file and it will require a communications link which is 25 times faster if the two are to be transmitted in the same time.

Most imaging applications will make use of LANs and in the early days of DIP many of these LANS were dedicated for use only by the DIP system. However, as bandwidths continue to expand, DIP facilities are offered as just one facility to run over a corporate network. Increasingly DIP is being offered over Wide Area Networks and the introduction of ISDN (Integrated Services Digital Network) will accommodate the requirement for a high bandwidth but this may be on a periodic basis only.

Because of the need to provide links with other corporate systems, the network architecture should be as open as possible and should support protocols such as X.25 or TCP/IP. Using such protocols in a UNIX environment it is relatively easy for systems integrators to connect PC compatible workstations with a UNIX storage processor.

8.12 Summary

As mentioned earlier, DIP is not one technology but rather a number of technologies which have been brought together to provide the imaging process, as described in the early part of this chapter. It is not intended that this publication should cover all the technologies in detail, but rather provide a general introduction and point readers in the direction of the *NCC's DIP Resource Pack* should they require additional in-depth material. From experience, so many companies become embroiled in the technology of DIP at the expense of non-technical issues, covered in Chapter 3. It is very easy to do but the success of your DIP installation depends, just as much, if not more, on understanding your information flows and what you want to achieve, as the precise details of how your scanner might work.

9

Conclusions and Future Developments

9.1 Introduction

This final chapter considers a number of technology trends which will have a direct effect on the use of DIP. In the second half of the chapter related technologies will be considered such as COLD (Computer Output to Laser Disk), EDI (Electronic Data Interchange) as well as ODA (Open Document Architecture) FTR (Full Text Retrieval,), all of which will impact on the wider information handling issue.

9.2 Technology Trends

The last few years has seen more and more computer capability available for less and less money and this trend is set to continue. This will obviously affect the price of DIP systems. It is predicted that within two years the price of a typical PC configuration will be reduced by a third and that the performance of such a system will have improved. The screen resolution will be superior and the typical 300 by 300 dpi printer will be replaced by a 400 dpi version.

In two or three years the 386 chip will be replaced by the 486 and it will not be long before the 586 chip which is 250 times more powerful than the original PC will have an even more dramatic impact. The cost of storage, based on a price per megabyte, will continue to decrease as the capacity of storage devices increases. It is predicted that the amount of information that can be stored on an optical disk will increase four-fold within four years. Compression algorithms are also becoming more powerful and therefore impacting on the storage needs. No longer will we be looking at a database of images but one that includes various forms of data including text, voice, images and video and possibly even in colour. These different forms of data will be handled in a form that is transparent to the user as well as the system architecture and software.

The multimedia database will make use of network and communications developments. Network capacities will increase and make it possible to transmit images over wider areas at an acceptable cost. The new Integrated Services Digital Network (ISDN) will provide combined voice and data networks which will allow the transmission of images via telephone lines.

Improvements in OCR technology (the term OCR being used in a generic sense to cover all character recognition technology) will allow the recognition of handwriting. This will impact on a number of manual processes associated with DIP and in particular the indexing of documents. DIP will also become more integrated with other applications. It will be just one facility within a wider office automation offering, or it could be linked with a project management application to control documentation such as that associated with the increased use of the BS 5750 standard.

What should also not be forgotten is that micrographics is still very much alive and in certain areas will continue to be used. It will not be left behind in isolation, as technology exists which makes it possible to bridge microfilm, magnetic, optical and paper stored images into a single database.

However, technology is not the only future development we should be concerned about. How the future workforce will make use of the technology advance possibly presents the greatest challenge. To make full use of the advances in technology, described earlier, will require a skilled workforce. Recent statements from governments claim that future school leavers do not have the same level of educational achievement as existing staff. In being able to capture, index and access information, the ability to read and write and understand English is of major importance.

Compounding this problem is the fact that the technology itself is getting more sophisticated to meet new demands and therefore it will become increasingly recognised that training will be the key to the successful use of future technology. This does not mean a two-day training course when new equipment arrives, but something far more comprehensive and detailed. It has been calculated that knowledge worker productivity is growing at less than 1 percent per year. Such people use technology where for example the processing capability of the equipment they use can rise by 30 percent per year. Looking at the state of technology advances which customers and their perceived needs have encouraged, perhaps the greatest challenge facing future uptake of DIP is to make future knowledge workers even more knowledgeable and capable.

9.3 Related Developments

Throughout this publication the idea has been put forward that DIP must not be seen as yet another island of automation but as part of a much wider move towards total information management. It may be easier to take a halfway approach at this stage and have as a mid-term objective a move towards electronic document management. Even small companies cannot isolate themselves from this issue. Increasingly companies are faced with the demands for electronic trading from major trading partners through the use of Electronic Data Interchange and the use of structured messages. In certain industry sectors to be sub-contractor for even the smallest component or service to the defence industry will shortly require compliance with the CALS standard.

9.3.1 COLD (Computer Output to Laser Disk)

Many companies have recognised for some time that paper was not the most suitable medium for storing computer-generated records and they turned instead to microfilm, microfiche, and magnetic tape for computer output storage. However, the information processing capacity of today's computer technology has begun to stretch the limits of such systems and increasingly data-intensive organisations are installing optical disk technology to store, access, and distribute computer records.

This technology is often referred to as COLD (Computer Output to Laser Disk) and the benefits of COLD are particularly clear for organisations that have already made an investment in optical technology. For companies already using imaging systems, COLD applications can be added at very low cost and can be used as an additional way of justifying their imaging investment because the payback for COLD technology is very fast.

COLD technology relies on the same storage medium as imaging technology and this can cause potential users some confusion. In fact, there are significant differences between imaging and COLD. A major difference relates to indexing and with COLD this process is fully automatic, and therefore, cheaper than imaging in which there is usually some manual intervention. The two systems also differ in terms of per-disk storage capacity. Using data compression techniques, COLD systems can store between one million and two million pages per disk; compared to the 40,000 – to 60,000-page capacity of most imaging systems.

Another major difference is the fact that instead of requiring an expensive workstation to view an image, information can be retrieved from a COLD system using a variety of hardware, including dumb terminals.

The process of putting computer output onto optical disk varies between vendors, but typically in a PC based system, industry standard intelligent file transfer is used to send the host computer data from the print spool directly to the hard drive of a PC. The software indexes and compresses the data and then files it in character-data format onto optical disk. The indices can be filed on either the PC's hard disk or the optical disk. Such systems can also be run over a local area network.

The capacity of most COLD systems can be expanded with library units, or jukeboxes in the same way as DIP systems.

9.3.2 ODA (Open Document Architecture)

This is not so much a technology development, rather an international standard which is of direct interest to the DIP arena. As well as being used in the exchange of documents between text processing systems with different hardware, software and incompatible data formats, this standard assists in the retention of corporate information resources. Over a period of time, documents prepared or stored on one system may become no longer accessible due to change or evolution of the equipment for accessing and manipulating documents.

The Open Document Architecture, ODA, is a new information interchange standard of major significance for use in Open Systems Interconnection. ODA was initially formulated in Europe by ECMA, the European Computer Manufacturer's Association, as their standard ECMA-101 (1985). Subsequently, ODA was adopted by ISO and then by CCITT as their single document architecture, to be applicable to all forms of telematic services involving electronic document interchange. ODA was originally known as Office Document Architecture.

The ODA Standard enables documents containing character text, raster graphic (facsimile) images and geometric graphics to be encoded and interchanged electronically.

9.3.3 Microform

The two components of microform, microfilm and microfiche, are the most cost-effective methods for storing inactive documents on a long-term basis. However, they are not appropriate for companies that cannot tolerate a long out-of-file time, or for situations where documents may need to be accessed for active processing or reference.

9.3.4 Electronic Data Interchange

Electronic Data Interchange (EDI) is primarily used to transmit and receive business documents, such as orders and invoices, between manufacturers and suppliers, according to predefined electronic formats. EDI works well in established customer-supplier relationships, in which both parties have approximately the same level of automation. The problem with EDI is that the great majority of business communication involves users who either cannot utilise EDI because of the incompatible software and standards, or who simply need more capabilities than EDI can provide. In these situations, imaging may be appropriate.

9.3.5 Full Text Retrieval (FTR)

The term Full Text Retrieval and free text retrieval are both used in connection with certain types of text retrieval systems. The two terms have come about due to slightly different needs when accessing different types of textual information. On the one hand there was structured information with fixed-length numeric and text fields in which, using certain search criteria whole records could be retrieved and on the other hand, the natural-language prose such as found in reports, periodicals and books.

The term Full Text Retrieval is now generally accepted as the generic title used to describe all such products which allow interrogation of fixed or free format textual documents. Use of FTR is obviously not possible in conjunction with images that are captured as bit-mapped files, but the inclusion of OCR and the capture of the image as an ASCII file offers much greater flexibility to what can be achieved. Case Study number 2, in Chapter 6, is a prime example of how FTR is used alongside DIP.

9.4 Conclusion

DIP is still a relatively new technology which to some companies can still seem prohibitive due to either cost, technical complexity or even because successful implementation may require a radical rethink in company processes and procedures. The pace of uptake is increasing, although not perhaps as quickly as many of the product providers, who see DIP as the technology topic of the 1990s, would like. As Frederick Wang said, "it will take time to realise its full potential", and this is in fact what is happening.

If you have come to the end of this introductory guide and your interest is sufficiently awakened to pursue the topic further, the book has achieved its objective, especially if there is a realisation that there are as many non-technical issues to be addressed as technical ones. DIP is not an end in itself, it is a beginning, a way of bringing the vast quantities of paper-based information into a corporate information management system. It is also the management of such a system, with information held as video, audio, image and in various other forms, that is the challenge for the next decade.

Appendix 1

Glossary

Address
The physical or logical location in a computer's memory of a quantity of data.

AM
See **Automated Mapping**

American National Standards Institute (ANSI)
The principal standards coordination body in the United States.

American Standard Code for Information Interchange (ASCII)
A widely used system for encoding letters, numerals, punctuation marks, and signs as binary numbers.

ANSI
See **American National Standards Institute**.

Aperture card
A Hollerith card (punch card) with a 35-millimetre frame of microfilm mounted in its centre. The microfilm is usually a picture of an engineering document, which can be as large as an E- or A0-size document.

Archival
Readable (and sometimes writeable) for a long time; long can mean from five years to more than 100 years.

ASCII
See **American Standard Code for Information Interchange**

Bandwidth
The range of frequencies that can be passed through a channel. A channel carrying digital information has a data rate proportional to its bandwidth. For videodisks the bandwidth is often 15 KHz on either side, or 30 KHz.

Bit-mapped display
A video monitor that stores bit patterns for digitised document images in

internal memory and selectively illuminates or darkens the display in areas that correspond to light and dark pixels in the original document.

Bits Per Inch (bpi)
A measurement of data density on a linear storage medium, eg magnetic tape.

Bubble
In optical memory, formations created by a laser in an optical recording medium. *See* **mark**.

Cache storage
Temporary storage for data to which access must be very quick. *See* **magnetic disk cache**.

Cartridge
An enclosure, generally of plastic, in which an optical medium is kept for protection; also called a cassette.

CCITT Group 3 and Group 4
International compression/decompression standards for black and white images specified by the Consultative Committee for International Telegraph and Telephone (CCITT); the international standard for facsimile communications.

Charge-Coupled Device (CCD)
A semiconductor device capable of both photodetection and memory that converts light to electronic impulses. One and two dimensional CCD arrays are used in scanners to perform the first stage in converting an image into digital data.

Clustering
The process of grouping similar items or features for storage on a given platter or series of platters.

COLD
See **Computer Output to Laser Disk**

COM
See **Computer Output Microfilm**

Compact Disk
A read-only optical disk available in formats for audio, data, and other information. The most commonly encountered size of a compact disk measures 4.¾" inches (120mm) in diameter.

Compact Disk Read Only Memory (CD-ROM)
The compact-disk format for computer data. Generally used for the storage of relatively unchanging data and/or images, such as archival files.

Compression
Conversion of a digital image to a lower number of bits for storage. Contrast with decompression.

Compression algorithms
Procedures according to which digital data sets representing images are compressed. Some of the most successful of these are contained in the CCITT Group 3 standards for digital facsimile.

Computer-aided Acquisition and Logistics Support (CALS)
A US Department of Defence initiative supporting the electronic interchange of engineering documents between contractors and government agencies.

Computer-Aided Design (CAD)
Software to assist in performing standard engineering and architecture design functions.

Computer-Assisted Retrieval (CAR)
Broadly speaking using computer memory to store indices pointing to information that cannot be stored cost-effectively in a computer, and using the search capability of the computer to retrieve the desired information, which may be stored on micrographics or other storage media.

Computer Output Microfilm (COM)
A system in which digital information is converted into an image on dry processed microfilm. *See* **Raster format Computer Output Microfilm**.

Computer Output to Laser Disk (COLD)
A system used to control the transfer of computer-generated output to optical disk for on-line or off-line storage.

Crater
A mark made by transverse displacement of the material of an optical recording medium. A crater is distinct from a hole in that a crater has a raised lip around its perimeter. *See* **mark.**

Cropping
The truncation of an image to fit into a viewing window.

Decompression
Reconstruction of a compressed image for display or printing. Contrast with compression.

De-skewing
The adjustment made to an image to compensate for physical distortions inherent in the system, or the adjustment made to an image to compensate for justification errors in scanning.

Digitiser
A scanner that converts paper documents or microfilm images to a stream of bits representing tonal variations in successively encountered picture elements.

Dithering
A scanning technique for digitising grey areas by intermixing black and white pixels.

Document
A hard-copy or electronic representation of information.

Document digitisation
The use of scanners to convert documents to digitally coded, electronic images suitable for magnetic and optical storage.

Document processor
A computer within an imaging system that manages database transactions and image storage space.

Document Storage Processor (DSP)
A computer consisting of a processor that manages the storage devices of an electronic imaging system. All database transactions are managed by the DSP.

Dots Per Inch (dpi)
A measurement of resolution, eg the number of pixels per inch on a workstation display monitor.

Editing
Inserting, deleting, or changing attribute or geometric elements to correct and/or update a model or database.

Engineering Document Management System (EDMS)
A computer-based system for the electronic capture, storage, retrieval, distribution, viewing, and editing of engineering documents.

Erasable
Capable of being rewritten, either after bulk erasure or spot erasure.

Erasable optical disks
A type of read/write optical disk that permits the deletion of information and the re-use of previously recorded disk areas.

Facilities Management (FM)
A specialised Geographic Information System (GIS) used for managing the spatial (map) data for a physical plant, building, or facility. The primary userbase is plant maintenance, utilities, and telecommunication companies.

Fax server
A device that supports datafax input and output to a computer system.

Flip
To present the other side of a document image for viewing on an image workstation display monitor.

Folder management system
A computer-based system that provides for the electronic capture, storage, distribution, annotation, display, and printing of images previously available only on paper or other hard-copy media. The emphasis of a folder management system is the file folder, which contains multiple documents of varying numbers of pages, and for which the tracking of workflow is important.

Geographic Information System (GIS)
A system of computer hardware, software, and procedures designed to support the capture, management, manipulation, analysis, modelling, and display of spatially referenced data for solving complex planning and management problems.

Graphics windowing
The ability to partition the viewing area of a graphics monitor to support the execution of multiple parts of the same application or multiple applications.

Grey scale
A series of achromatic tones (tones with brightness but no hue) having varying proportions of white and black, to give a full range of greys between white and black. In computer graphics systems with a monochromatic (black and white) display, variations in brightness level (grey scale) are employed to enhance the contrast among various design elements.

Image
The digital representation of one side of a document or to form a digital representation of a document.

Image board
A hardware performance accelerator that, when added to a workstation, supports the rapid compression, decompression, and manipulation of images.

Image capture
A series of operations required to encode documentation in a computer-readable digitised form.

Image header
A part of an image packet with information to correlate the image data to document information stored in the host and to allow later retrieval of images. *See* **image packet.**

Image packet
A block of compressed image data prepared for transmission. Each packet holds data for a certain number of documents and contains information to correlate the image data to document information stored in the host computer. Image packets are buffered to allow for short-term variations in the transmission rate. *See* **image header.**

Image workstation
A computer workstation that lets the user manipulate imaged information.

Indexing
A method by which a series of attributes are used to uniquely define an imaged document so that it may later be identified and retrieved.

International Graphics Exchange Standard (IGES)
A standard designed to support CAD-based information.

Item
Any piece of paper that can be processed by a document processor.

Journal drive
An optical disk drive that records all document database storage transactions, including image and attribute data.

Jukebox
An automatic selection and retrieval device that provides rapid, on-line access to multiple optical disks.

Laser
Light Amplification by Stimulated Emission of Radiation. A device for generating coherent radiation in the visible, ultraviolet, and infrared portions of the electromagnetic spectrum.

Magnetic disk
A flexible or hard-disk medium used to store data in the form of minute local variations in magnetisation of the disk surface.

Magnetic disk cache
A directory on magnetic disk in the Document Storage Processor (DSP) that provides storage for, and quick access to, frequently used documents.

Magnetic tape
Ribbons of film on which data can be stored in the form of minute local variations in magnetisation of the iron compounds in the film.

Magneto-optical disk
A type of erasable optical disk that uses laser optics to alter the polarisation of a magnetic disk coated with iron in combination with rare-earth transition metals.

Mark
A hole, pit, crater, bubble, or other irregularity that a laser makes on an optical medium.

Markup
The imaging system function that supports redlining or creating modifications to a document. Changes to a document are created in a file that acts as a transparency superimposed on the original document; the original is not altered.

Master
An original recording, in its final format, intended for mass replication.

Microfiche
A technique for storing multiple pages of a document on a single sheet of photographic film.

Microfilm
The processed photographic film kept for later retrieval and viewing for research purposes.

Micrographics
A generic term that encompasses microfiche, microfilm, rolled film, aperture card, and similar technologies.

Open Systems Interconnection (OSI)
A mass storage interface standard promulgated by the International Standards Organisation (ISO)

Optical card
Wallet-size, plastic data storage medium coated with an optical recording material, usually in a stripe.

Optical Character Recognition (OCR)
A device that scans printed documents and attempts to recognise the letters, numerals and other characters the documents contain, and convert them to ASCII representation for computer storage.

Optical disk
Platter-shaped medium coated with an optical recording material.

Optical filing system
A computer-based hardware and software configuration that stores digitised document images on optical disks for on-demand retrieval.

Optical storage
Technologies, equipment, and media that use light – specifically light generated by lasers – to record and/or retrieve information.

Optical tape
Ribbons of film for data storage, coated with an optical recording material and kept in reels, cartridges, or cassettes. Compare with magnetic tape.

Picture elements
Areas into which a document is divided for scanning purposes, often abbreviated as pixels or pels.

Platter
A large, round disk for storing information, eg a phonograph record. *See* **optical disk.**

Plotter
An output device for drawing maps or engineering drawings.

Prefetching
The process of building a queue of images for subsequent processing.

Raster
The array of scan lines used to cover a planar area to read or depict image information on that area. The electron beam of a TV picture tube writes a raster on the phosphor screen. Compare with vector format.

Read-only optical disk
An optical disk produced by mastering processes, containing pre-recorded information.

Reduction rate
The degree to which a document is compressed, and the associated space savings in memory.

Resolution
As applied to document scanning, the specific pattern and number of picture elements sampled by a given scanner, or displayed by a graphics monitor. Resolution indicates the potential for detail in scanned images and is an important determinant of image quality.

Rotate
To change the orientation of information (such as images) displayed on a workstation monitor. For example, a 90-degree rotation allows a vertical object to be viewed in a horizontal orientation.

Scaling
A technique that reduces or enlarges an image by combining picture elements (pixels). For example, a two-by-two pixel representation can be combined by various techniques into one pixel.

Scanner
A device that resolves a two-dimensional object, such as a business document, into a stream of bits by raster scanning and quantisation.

Scan service/device
The input device used to enter documents into an imaging system. These devices run a hard-copy document under an image sensor. A scanner interprets the reflected light from the hard-copy document and converts the information into digitised raster data.

Scroll
The action of moving images or text forward or backward (up or down) on an image workstation monitor.

Sensitive layer
The layer in an optical medium on which information is recorded; it may be composed of more than one layer. Materials include many plastics, semiconductors, and metals.

Skewing
An image condition resulting from physical distortion inherent to a monitor or from justification errors in document scanning.

Speckling
Extraneous dots that appear on an imaged document. Speckling may result from the presence of dirt on the hard copy or from an improper scanner setting.

Tag Image File Format (TIFF)
A de facto standard file format designed to promote the interchange of digital image data; developed jointly by Aldus and Microsoft.

Threshold
A setting, generally used in scanning, that determines whether a particular section of the document is white or black; that is, a threshold value that decides the white-to-black transition within a grey level.

Track
A linear or circular path on which data is encoded on an optical disk.

Turnkey optical filing system
A preconfigured combination of computer and optical storage components, including both hardware and software, which is purchased as a self-contained unit and is designed specifically for automated document storage and retrieval.

Vector format
A representation of a line drawing by listing the beginning and end points of all the lines.

Videodisk
An optical disk that stores information in the form of standard television signals. Videodisks can store either analog or digital information. Most videodisks are read-only media, but several companies offer videodisk records that employ write-once optical disks.

View
The imaging system function that lets the user look at stored documents and overlays.

Workflow management
The ability to route and track electronic documents through an organisation in a procedural fashion.

WORM
See **Write Once Read Many times.**

Wrap
A compression technique that ignores line boundaries, resulting in greater document compression ratios than achieved by non-wrap techniques. Contrast with non-wrap.

Write Once Read Many times (WORM)
An attribute of certain optical disks. Once information is 'burned' into a sector of the disk it cannot be deleted or overwritten, but can be read any number of times.

Zoom
To enlarge the presentation size of information (such as text or images) displayed on a workstation monitor.

Appendix 2

Information on typical DIP systems and suppliers

ADVENT SYSTEMS (IMAGING) LTD

Company Name:	Advent Systems (Imaging) Ltd
Address:	12 The Business Centre Molly Millar's Lane, Wokingham Berkshire, RG12 2QZ
Telephone Number:	0734 784211
Sales Contact:	Clare Barrett
Company Background:	Advent has been in the business of Document Image Processing since 1984 when the company was founded. Advent is a British Company and has distributors in Europe, USA and Australia.
Target Markets:	Business Document Management; Engineering Drawing and Document Management; Geographical Information Systems ASARI (Drawing Office Document Management System) provides the facility to scan, edit, store, retrieve and plot historical drawing data from A4 to A0. The system comprises full raster and vector editing and redlining capabilities together with a Document and Drawing Management System. This allows other

relevant documents such as test certificates and manuals to be linked to the drawing. Documents can also be viewed on PCs.

ASAR2 (Business Document Management) is a fully user-definable system enabling documents to be scanned at high speed and then indexed, stored and retrieved. It uses a graphical user interface providing an easy to use system. Other modules include browsing, fax transmission, folders, printing and mailing tools.

ASAR4 (Design of Low Voltage Distribution Networks) providing the electrical engineer with a graphical user interface for the design of low voltage distribution networks, using raster maps or building plans as the basis for the interactive design of a network.

CACL

Company Name:	Computer & Aerospace Components Ltd
Address:	3 Kingston Business Centre Fullers Way South Chessington Surrey KT9 1HW
Telephone Number:	081 397 5311
Facsimile:	081 391 4835
Sales Contact:	D R Abbott
Target Markets:	Office Automation Research; Litigation Support; Libraries;
	PCs, Networks; DOS, Novell, Token Ring;
	Systems sold as 'add-ons' to existing PC/laser installations.
	Image compression/decompression in software providing low-cost networking.
	Wide variety of displays/optical disks/scanning systems supported.

CALLHAVEN PLC

Company Name: Callhaven plc

Address: 74 Rivington Street
London
EC2A 3AY

Telephone Number: 071 410 9191

Facsimile: 071 410 9195/6

Product Offerings: Callhaven are the sole UK Distributors of the leading DIP product in the US Market MD MARS (Multi-User Archival and Retrieval System). They also supply an associated Full Text Retrieval product Freeform which may be used in conjunction with the MARS Imaging engine.

The flexibility of the MARS imaging engine has meant that it has solved problems in a wide variety of application areas. Current systems are performing functions as diverse as a duplicate billing systems for a telephone company with millions of clients. Oil well drilling logs, litigation support, medical records. University administration systems, Legal documents and contract archive, Correspondence administration, on-line newspaper cutting systems, Signature verification in a bank to photographic libraries. In addition it is used in various government departments including the US Navy, Army, Air Force and NASA.

CIMAGE INTERNATIONAL LIMITED

Company Name:	Cimage International Limited
Address:	Centennial Court East-Hampstead Road Bracknell Berkshire RG12 1JZ
Telephone Number:	(0344) 860055
Facsimile:	(0344) 861471
Contact:	Mike Tarttelin (Marketing Director)
Products:	ImageMaster – Modules for the capture, QC, editing, manipulation, viewing and markup of technical documentation: Drawings, specifications, maps, plans, certificates, layouts, etc. Document Manager – Modules for the management and control of technical documentation: change control, access control, revision control, Markup management, document linking, distribution, etc.
Category:	Technical Document Image Management System (TDIMS)
Business Focus:	Cimage focus exclusively on providing applications that solve the document management problems faced by manufacturing, construction and engineering services organisations. It develops and integrates computer based image management products that capture, edit, manage and distribute technical and engineering documents cost-effectively. These products improve the quality of technical documents – and the operational efficiency and effectiveness with which organisations manage document control and distribution.

ELECTRONIC DOCUMENT SYSTEMS LTD

Company Name:	Electronic Document Systems Ltd
Address:	Christy Estate Ivy Road Aldershot Hampshire GU12 4TX
Telephone Number:	0252 316141
Sales Contact:	Mr Tom Bird
Background:	Electronic Document Systems was established in 1989 to address the corporate and end-user market for PC-based Document Image Processing (DIP) systems.

It has installed DIP systems in a wide range of commercial applications for customers throughout the country including British Sugar plc, Chase Manhattan Bank, Government establishments and a number of local authorities.

The company address a wide range of commercial applications for the storage of paper-based documents including purchase order documentation, insurance documentation, general customer information and technical manuals up to A3 in size. In particular it has expertise in the banking/insurance and distribution sectors.

All its systems are PC-based and operate generally under the popular MS-Windows environment.

IBM LTD

Company Name: IBM Ltd

Image Products: IBM UK announced the first ImagePlus products in March 1989. Since then more than 200 customers have installed Operational Image applications worldwide.

There are two ImagePlus products:

- SAA ImagePlus MVS/ESA for users of ES/9000, S/390 or S/370 systems. Installations range in size from 15 to 1500 image workstations. There are several in the range from 200 to 700 users.
- SAA ImagePlus/400 for users of AS/400 systems. Installations range in size from 2 to 500 image workstations. Many are in the range from 10 to 50 users.

Both products required IBM/PS/2 workstations for the capture and display of images – which are indexed, stored and distributed by the host system.

ImagePlus provides folder management, case processing and workflow management in addition to storage and retrieval of large volumes of images. It is useful therefore for a very wide variety of industries and customers. Examples of its uses: Accounts payable; Investment portfolios; Legal case files; Customer service, Waybill processing; Personnel; and Insurance (claims/new business/archiving), as well as established applications.

ICL PLC

Company Name:	ICL PLC
Address:	Observatory House Slough SL1 2EY
Telephone Number:	0753 516000
Sales Contact:	Martin Spencer
Background:	ICL has been active in DIP since 1983. Prototypes of DIP solutions were run on three customer sites to provide feedback for the development of POWERVISION, launched in October 1990. POWERVISION is ICL's first DIP product to be announced, providing File Server and Image Workstation Management software and an Application Programming Interface to C library subroutines. The logical index can therefore be provided by many database products. POWERVISION provides Document Image Processing support of office documents in departmental and enterprise-wide applications.

INTEGRATED DOCUMENTS LIMITED

Company Name: Integrated Documatics Limited

Address: Olympic House
196–200 The Broadway
Wimbledon
London
SW19 1SL

Telephone Number: 081 543 3611

Facsimile: 081 542 4415

Sales Contact: Christopher Roffey, Business Development Manager

Background: Target market sectors are Financial Services (banking and insurance), Government (local and central), and Energy. Other include Health, Pharmaceuticals, Aerospace, Manufacturing and Food and Drink.

Applications: Include Finance Sector – customer files, credit files, underwriting, claims loans, mortgages, direct debit mandates, signature verification. Government Sector – community charge, housing benefits, planning, payroll files, personnel records, contract administration, case management.

Energy Sector – project records, maintenance records, operations data, drawing control, certification data, permit to work, QA safety documentation.

IMAGESOLVE INTERNATIONAL PLC

Company Name:	Imagesolve International plc
Address:	4 Dollis Park Finchley London N3 1HG
Telephone Number:	081 346 0247
Facsimile:	081 346 1447
Sales Contact:	Helen Parslow, Marketing Manager
Background:	One of the major reasons claimed for its succes has been its ability to supply application specific systems based upon standard and freely available PC/AT, PS/2 or compatible components. The Imagesolve family of products incorporates both data storage to optical disk and paper documents. The product mix comprises a range of systems from a very simple index database, to Imagesolve II which runs under Microsoft Windows and SQL, providing co-operative processing with mainframe database environments. The family of products includes systems designed for unique needs, such as signature verification and data storage.

IMAGE SYSTEMS EUROPE LTD

Company Name:	Image System Europe Ltd
Address:	Sheffield technology Park 60 Shirland Lane Sheffield S9 3SP
Telephone Number:	0742 420419
Facsimile:	0742 445197
Sales Contact:	Denise Stott, Ken George
Background:	Image System Europe Ltd was formed in January, 1990, by a group previously operating as the image system group within British Steel's central management services division. Image Systems Europe enjoys complete independent of platform and software. DIP applications are provided using industry standards of technology, design and implementation to provide harmonious integration with existing and proposed information management systems.

IRIS SOLUTIONS LTD

Company Name:	Iris Solutions Ltd
Address:	Genesis Centre Science Park South Birchwood Warrington WA3 7BH
Telephone Number:	0925 852220
Facsimile:	0925 852223
Sales Contact:	Steve Kaye
Background:	Iris Solutions was established in 1990. Is UK distributor for the ODIN DIP software produced by GN Filetch of Denmark and is one of Hewlett Packards Premier Solutions Provider. Target markets are technical and commercial and its speciality is the integration of DIP applications with existing management and technical systems.

OLIVETTI SYSTEMS AND NETWORKS UK

Company Name: Olivetti Systems & Networks UK

Address: Image Processing Technologies
86–88 Upper Richmond Road
Putney
London
SW15 2UR

Telephone Number: 081 785 6666

Sales Contact: Carol Curtis, Marketing Department, Image Processing Technologies

Background: Image Processing Technologies Division

The Image Processing Technologies Division was established in 1986, with the first sale of a DIP system to Britannia Building Society.

FileNet

Manufactures and markets the FileNet DIP system from its headquarters in Costa Mesa, California. Olivetti has distributed the product worldwide for FileNet since it was set up in 1982 and both parties continue to work together cooperatively in maximising opportunities both in the UK and elsewhere.

Its target markets are Major Accounts, Finance, Insurance, Government and Retail.

PHILIPS INFORMATION SYSTEMS

Company Name:	Philips Information Systems
Address:	Elektra House Bergholt Road Colchester CO4 5BE
Telephone Number:	0206 575115
Facsimile:	0206 562140
Contacts:	Tony Moss, Sales
Background:	Philips Information Systems is part of the Worldwide Philips organisation and, in particular, is the UK sales arm of Philips Information Systems headquarters in Holland. Philips has been in the business of installing DIP systems since 1983, when it installed the first European DIP system at Grünher and Jahr in Germany. Since that date, Philips has installed over 250 DIP multistation systems. In the UK, Philips has installed systems mainly in the Finance/Insurance Sector, but also in Local Government.

STANDARD PLATFORMS

Company Name:	Standard Platform Ltd
Address:	Research & Development Centre Glenfield Park Northrop Avenue Blackburn Lancs BB1 5QF
Telephone:	0254 682442
Facsimile:	0254 583003
Sales & Marketing Centre	130 New Street Andover Hants SP10 1DR
Telephone:	0264 333999
Facsimile:	0264 333873
Sales Contact:	Brian Hamilton – Sales & Marketing Director

Background:

The company is a systems house specialising in the design and development of document processing and mass storage systems. Standard Platforms is able to combine its expertise in optical storage technology with an in-depth understanding of the UNIX operating system and the latest data communication techniques. It is this combination of hardware and software engineering skills, together with broad-based systems integration experience, which it is claimed, sets the company apart from others in its field.

It has a wide DIP target market.

UNISYS LTD

Company Name: Unisys Ltd

Address: Stonebridge Park
London
NW10 8LS

Telephone: 081 965 0511

Facsimile: 081 961 2252

Contact: John Lewis – Programme Director InfoImage Products

Unisys has many years of experience in the use of imaging technology, particularly in the field of cheque processing. In 1989 the Info-Image range of image-enabled applications was launched, including systems for Engineering Drawings Management, File/Folder management, ARGIS (Advanced Relational Geographic information systems) and IPS (Item Processing System)

Target markets include Aerospace, Automotive, Defence, Gas, Electricity, Water, Telecomms, Insurance, Banking, Building Societies, Government, Pharmaceutical.

WANG (UK) LTD

Company Name:	Wang (UK) Limited
Address:	Wang House 1000 Great West Road Brentford Middlesex TW8 9HL
Telephone Number:	081 568 9200
Facsimile:	081 847 1352
Contact:	David Allcock, Marketing Manager Imaging and Voice
Background:	Wang Laboratories Inc, based in Lowell, Massachusetts, designs, manufactures, and markets computer systems and provides related products and services for the worldwide information processing marketplace. It provides imaging capability within a range of environments, from PC through to IBM CICS and IMS. Their target markets are Banking, Insurance, Professions, Government, Manufacturing, Health Care.

Appendix 3

Further Reading

Additional NCC DIP Publications

Detailed below are two additional NCC Publications which will be of interest to readers should they wish to take the topic of DIP further.

DIP Applications Handbook edited by NCC

This new handbook is the latest and most comprehensive guide for aspiring users of Document Image Processing, containing the largest set of DIP case studies ever assembled.

It provides prospective users with a detailed assessment of the market, including revealing information from DIP suppliers but also offering an impartial overview of this highly-volatile marketplace. The full range of DIP applications are considered, together with a review of the wide variety of solutions in terms of price and performance.

While describing many of the technical aspects, the handbook gives equal emphasis to the business and implementation issues, thus providing a balanced approach (regardless of company size) for those planning to use DIP.

The Handbook provides:

- case studies – detailing the application area, the system solution and the facilities provided in over 70 real-life applications;
- supplier profiles – listings of the hardware, software, target markets and particular specialities of over 40 major DIP suppliers;
- extensive user listing – details of user sites in the UK, USA and Europe, indexed to allow easy reference.

Major suppliers listed include: Advent Systems, Bell & Howell, Bull, CACL, Hewlett Packard Ltd, IBM, ICL, Kodak Ltd, Olivetti Systems, Scan Optics Ltd, Wang, Philips Information Systems.

Set of three DIP Reference Volumes

A survey by NCC identified that 61 percent of companies expect to be using DIP within two to five years. Despite this high level of interest, respondents also identified three major areas of concern: the complexity of DIP technology, the viability of its application, and its cost.

Those are precisely the issues addressed by these three reference volumes, in the NCC DIP library, making their publication unique and timely. Since each is based on wide, practical experience of DIP application and implementation, they are also uniquely relevant to the real needs of anyone considering – or advising on – the use of DIP.

The three volumes are:

Technical Introduction – an in-depth guide for the technical specialist responsible for implementing DIP systems and integrating them with existing applications.

DIP Case Studies from Europe and USA – documenting practical user experience to demonstrate why they chose DIP to solve their business problems, and the benefits already achieved.

The Business Benefits and Justification – the management guide to the cost, functional, and corporate benefits of DIP and techniques for its business-related justification.

The three volumes, comprising nearly 600 pages of information, offer a uniquely authoritative and balanced guide. Together they provide a comprehensive resource to satisfy the requirements of managers and technical staff, in every organisation which needs to know how? when? and where? to exploit the potential benefits of DIP.

Further information on these and other NCC publications can be obtained from: The National Computing Centre Limited, Oxford Road, Manchester, M1 7ED. Tel: 061 228 6333.

Appendix 4

Useful Addresses and Contacts

Listed below are brief details and contact information for organisations and user groups that may be able to provide additional information on DIP, as well as details on a number of DIP magazines.

CIMTECH

Set up by Hatfield Polytechnic, CIMTECH provides independent advice on a wide range of information-handling topics, from microform through to image processing.

Address: CIMTECH
Hatfield Polytechnic
College Lane
Hatfield
Hertfordshire
AL10 9AD

Telephone: 0707 279691

Document Image Automation

An American magazine published every two months by Meckler which covers a wide spectrum of document image applications.

Address: Meckler Corporation
11, Ferry Lane West
Westport,
CT 06880
USA

Telephone: (203) 226–6967

Electronic Document Management Circle

The Electronic Document Management Circle provides a forum for the exchange of opinions, experience and information between users, suppliers and independent experts.

Address: The National Computing Centre Limited
Oxford Road
Manchester
M1 7ED

Telephone: 061 228 6333

FINTECH – Electronic Office

A bi-monthly newsletter published by the Financial Times which covers the area of the electronic office and makes considerable reference to DIP activities.

Address: Financial Times Newsletters,
Tower House,
Southampton Street
London WC2E 7HA

Telephone: 071 240 9391

Image Processing

A UK publication which covers a range of imaging topics from medical and scientific through to DIP.

Address: Image processing
Reed Business Publishing Ltd
Oakfield House
Perrymount Road
Haywards Heath
Sussex RH16 3DH

IMC Journal

This journal of the International Information Management Congress is published every two months and covers a range of DIP and microform topics.

Address: IMC Journal
345 Woodcliff Drive
Fairport
NY 14450
USA

Telephone: (716) 383 8330

INFORM

The Magazine of the Association for Information and Image Management in America.

Address: 1100 Wayne Avenue,
Suite 1100,
Silver Spring,
MD 20910,
USA

The National Computing Centre Limited

A source of independent information and advice on DIP providing a range of publications and consultancy services

Address: The National Computing Centre Ltd
Oxford Road
Manchester
M1 7ED

Telephone: 061 228 6333

UK AIIM (Association for Information and Image Management)

A group of users and suppliers with interest in both information and image management including microfilm.

Address: c/o CIMTECH
Hatfield Polytechnic
College Lane
Hatfield
Hertfordshire
AL10 9AD

Telephone: 0707 279691.

Index